THE SURFACE FEATURES OF THE LAND

ALFRED HETTNER
The Surface Features of the Land

PROBLEMS AND METHODS OF
GEOMORPHOLOGY

Translated and with a preface by
Philip Tilley

MACMILLAN

This translation from the German of the second edition
of DIE OBERFLÄCHENFORMEN DES FESTLANDES
published by B. G. Teubner Verlag, Stuttgart, in 1928.

© Translation 1972, Philip Tilley
© Preface 1972, Philip Tilley

All rights reserved. No part of this publication may be reproduced or
transmitted, in any form or by any means, without permission.

First published 1972 by
THE MACMILLAN PRESS LTD
London and Basingstoke
Associated companies in New York Toronto
Dublin Melbourne Johannesburg and Madras

SBN 333 00698 4

Printed in Great Britain by
BUTLER & TANNER LTD
Frome and London

Contents

TRANSLATOR'S PREFACE	ix
AUTHOR'S FOREWORDS TO THE FIRST AND SECOND EDITIONS	xxi
INTRODUCTION	xxiii
The importance of geomorphology	xxiii
Geomorphology and its media	xxvi
The doctrines and methods of W. M. Davis	xxx
The aim of the book	xxxi
CHAPTER 1: THE MINOR FEATURES OF THE LANDSCAPE	1
The importance of minor landforms	1
Methods of description and investigation	3
Classification and terminology of minor landforms	5
The action of forces and the analysis of landforms	5
The dependence of landforms on rock-type and disposition	8
The dependence of minor landforms upon climate	10
Minor landforms of the past	11
CHAPTER 2: THE ORIGIN OF VALLEYS	14
The erosional nature of valleys	14
The theory of erosion	18
Direct, indirect, vertical and lateral erosion	20
The profile of equilibrium	22
River and valley meanders	25
CHAPTER 3: VALLEY TERRACES	27
The different classes of valley terraces	28
Gravel terraces	32

CHAPTER 4: AGE AND FORM OF VALLEYS ... 35

Age and stage of development ... 35
Davis's interpretation of age ... 37
Age characteristics ... 40
The psychology of error in ascertaining age characteristics ... 45
Canyons and other valley forms ... 47

CHAPTER 5: ALIGNMENT AND ARRANGEMENT OF VALLEYS ... 52

Consonant, inconsonant, surviving and subsequent valleys ... 52
Origins of valley alignment ... 57

CHAPTER 6: BENCHLANDS, REMNANT SURFACES, AND OTHER PLANATIONS ... 60

Benchlands ... 61
Other present-day planations ... 68
The theory of remnant surfaces ... 70
The occurrence of remnant surfaces ... 77
Criteria for the reconstruction of remnant surfaces ... 80

CHAPTER 7: STRUCTURAL PLAN AND STRUCTURAL STYLE OF MOUNTAINS ... 85

The structural plan ... 85
The structural style ... 88

CHAPTER 8: DEPENDENCE OF THE LAND SURFACE ON INTERNAL BUILD ... 92

Geomorphology and rock formation ... 93
A concept of age ... 97

CHAPTER 9: THE DEVELOPMENT OF THE LAND SURFACE ... 100

The nature of development ... 100
Tectonic development ... 102
Climatic development ... 107

CHAPTER 10: THE GEOMORPHOLOGICAL INTERRELATIONSHIPS OF LANDSCAPES 115

Processes by which material is moved 115
Fluvial transference of material 119

CHAPTER 11: COASTS 124

The study of coastline processes 125

CHAPTER 12: THEORIES ON THE ORIGIN OF THE LAND SURFACE 129

The scientific development of geomorphology 130
Davis's 'cycle' theory 134

CHAPTER 13: LANDFORM ASSEMBLAGES 137

Classification 137
The morphological character of landscapes 140
Classification of surface modification 143

CHAPTER 14: SUBDIVISION AND GROUPING OF THE LAND SURFACE 145

APPENDIX: GEOMORPHOLOGICAL RESEARCH AND PRESENTATION 148

Methods of geomorphological research: comparative map study 148
Methods of geomorphological research: induction and deduction 149
Methods of presentation 153
Terminology 156
Orometry 160
Morphological maps 161
Illustrations and views 163

GEOLOGICAL TABLE FOR SOUTH GERMANY 166

AUTHOR'S AND TRANSLATOR'S NOTES AND
    BIBLIOGRAPHIC REFERENCES   167

AUTHOR'S NOTES AND BIBLIOGRAPHIC
    REFERENCES   169

TRANSLATOR'S NOTES AND BIBLIOGRAPHIC
    REFERENCES   176

INDEX   191

Translator's Preface

However small a part of his field a geographer chooses to review and research, he has to take account of a very diverse and ever-growing mass of relevant literature. Faced with this he may easily overlook even the most noteworthy study in a language other than his own, and when he does, add to the already scandalous lack of communication and common intellectual ground between scholars who profess the same discipline. There is, therefore, an obvious and undeniable value in translating, and so making more readily available, a work considered to be representative of current thinking or past achievement.

But Alfred Hettner's *Die Oberflächenformen des Festlandes* is not such a work. First published in 1921, the book has not been republished in any language since a second German edition was issued in 1928. It is neither a modern and immediately useful study in geomorphology, nor a generally recognised classic of permanent interest or historical importance. Why, then, translate the book into English and republish it now? Not because there is a demand for it from English-speaking geographers, but because it may now be opportune for us to do what Hettner induced his German colleagues to do half a century ago: review the present state of landform studies within our discipline, and in particular assess the work of W. M. Davis as a contribution to landform geography. Hettner's book may help us to do this.

From 1884, for almost fifty years, Davis repeatedly expounded his scheme for the 'explanatory description' of terrain in terms of its evolution as a system of areally interrelated landforms. And he did this so persuasively that it was not long before English-speaking geographers, even specialist geomorphologists, could hardly conceive of any alternative approach to landform study. The scheme's usefulness in teaching, and the unity its widespread adoption brought with it, confirmed geographers in a view to which they were only too ready to hold on: that present-day landforms are best understood, or indeed can only be understood, if we try to reconstruct the stages through which they have passed in their development. But more recent and more exact study of the landform-shaping processes has shown that there is not necessarily any close connection between the developmental history of a landform and its present functional significance, in an areally organised system. As a result, geomorphologists have become increasingly aware that the genetic study of landforms is only one of the approaches open to them. Instructive as the study of landform development can be, it has to be founded on the direct and detailed investigation of the processes which shape landforms, and through which landforms are interrelated both in time and space. Yet neither landform history nor the study of landform dynamics can take the place of landform geography. The characterisation and comparative study of the diverse areal systems manifesting the interaction of landforms with one another, and with other kinds of earthly phenomena, can improve our understanding of them in a way no other approach can.

As Davis's overwhelming influence disappears, and with it the restrictive idea that the best way to understand landforms is to understand their evolution, geographers now have a chance to reassert the value of their discipline. What is more, they need to do so. Having given up the theory of the cycle of erosion as inadequate, and so lost the conceptual foundation that theory gave them, many English-speaking geomorphologists are inclined to look for another no less comprehensive scheme. And they, like Davis, have good reason to favour a scheme that is both useful in teaching and in harmony with current scientific attitudes. In such a situation geographers must be careful or the contribution their discipline can make to the understanding of landforms will again be overlooked.

Though at first sight it is no more than the outcome of Hettner's strong reaction to Davis's growing influence among German-speaking geographers at the turn of the century, *Die Oberflächenformen des Festlandes* is in fact rooted in its author's long and painstaking study of the nature of geography and its place among the sciences. Since the results of this study had been available to Hettner's German colleagues for almost twenty years when the book was first published, and because the book was largely a timely and well-considered restatement of views Hettner had maintained for at least ten years previously, it had an immediate and lasting effect on all German geographers before it was discarded. In Germany, as elsewhere during the latter years of the nineteenth century, the study of landforms was developed into a recognised field of higher learning, hand in hand with geography. In other countries there was an attempt to make the young science of 'geomorphology' serve what had come to be seen as the proper goal of geography, namely an understanding of the relationships between the natural and human phenomena of the earth's surface. In Germany, however, it remained free to go its own way towards becoming an autonomous earth science. It did so largely as a geography of landforms.

From the middle of the nineteenth century it was open to German students of landform to develop their field as a purely descriptive and classificatory account. To achieve progress in this direction they need only have developed the morphographic and morphometric appraisal of landforms begun by Humboldt and Ritter, and so never have gained more than a speculative understanding of how landforms originate. In seeking an explanatory understanding, inspired in particular by Richthofen, German scholars were inclined to pay as much attention to the forces and processes which shape the land surface as to the form of the surface itself. And when influenced by Davis, and somewhat later by Walther Penck, German geomorphology could easily have become an essentially geological discipline, and used its study of the present configuration of the land surface to reconstruct the stages through which this had developed. In fact, as Schmitthenner has shown,† German geomorphology was developed in the closing years of the nineteenth century as a geographical rather than a geological or geophysical discipline. The field gained its acknowledged place among the earth sciences in the German-speaking countries primarily as a comparative study of the distinctive areal assemblages of landforms that characterise different parts of the earth surface, and distinguish them from other parts. Moreover, we can now see that it was largely through their development of geomorphology as a geographical discipline that German scholars safeguarded and developed geography when its status as a field of learning, distinct from kindred earth sciences, was generally in doubt. The progress of landform geography was inevitably the concern of all German geographers.

The year before Davis's first statements of the cycle of erosion, Richthofen had set out, in his inaugural address at Leipzig,[b] the general direction German geography would take to remain consistent with the tradition consolidated by Ritter and Humboldt. Yet, as a

† See translator's notes at the end of this Preface, pp. xix.

geographer faced with proposals such as Gerland's that geography should be concerned only with the natural, non-human features of the earth surface, and with the forces that produce these, Richthofen was no more certain than most of his colleagues how to develop geomorphology as a geographical discipline. And in 1903, after twenty years of effort, he still could not reconcile the worldwide study of typical landforms with that of individual, localised ones and the way these are grouped together in areas. Despite the general agreement German geographers reached between 1875 and 1895 on the essentials of their discipline, they had not been able to exemplify this in the way they studied particular kinds of phenomena, for example landforms. As a result, German geography still had no assured place among the sciences. It is not surprising, therefore, that about this time German geographers began to see the concept of the cycle of erosion, and the scheme Davis had built around it, as a way out of their difficulties. For by the time Davis's scheme became widely known outside the United States it was no longer presented by its author as merely a *method* which could be profitably used in geological investigations[c], by the turn of the century it had become the only way to study the supposed reality of *The Geographical Cycle* of landform development.[d] Henceforward, Davis propounded his scheme as 'a well-coordinated system of classification' that would enable geographers to relate the inorganic landforms of the physical environment to the organisms, including man, which they carried.[e] Thus, Davis argued, the great variety of phenomena studied by geography in common with fields such as physics, chemistry, and history, could be integrated in a disciplined way to give geography the scientific standing it lacked compared with, say, the biological sciences. Attracted both by the systematic and comprehensive way in which the scheme related rock type and disposition with surface forms, and by the prospect the scheme opened up of a unified treatment of human and non-human phenomena, German geographers led by Albrecht Penck began to adopt it as a conceptual structure for their landform studies.

In the same volume of the *Geographische Zeitschrift* in which he published Albrecht Penck's exposition of 'physiogeography', with its approving references to Davis's scheme,[f] Hettner published his own most important study on the nature of geography.[g] From this study it is evident that Hettner was bound to see Davis's growing influence in Germany not only as a threat to the geographical study of landforms but also to the character geography had acquired in the course of its long history, and the distinctive place this gave it as a field of learning. Whatever Davis's original aims may have been, his system of landform study had become inextricably coupled with the idea that geography could be identified as a science of relationships, and specifically as the science most concerned with studying the 'correlation between environed organisms and the physical environment'.[h] With the publication of a German version of Davis's textbook *Physical Geography*, and of the lectures he gave at the University of Berlin in 1908–9,[i] Hettner was impelled to defend the place geography had established for itself as a chorological science, as a body of learning gained through the comparative study of areas. He sought to do this by showing how inconsistent Davis's approach was with geomorphology's tradition as a geographical discipline, whatever it contributed to our knowledge of landforms. Thus it may not be too much to claim that it was largely through his methodological writings on landform geography, and especially *Die Oberflächenformen des Festlandes*, that Hettner helped German geographers to a greater degree of agreement and mutual understanding about the nature and practice of their discipline than has so far been achieved elsewhere.

Regrettably, the series of hypercritical reviews of Davis's work that Hettner published from 1911 onwards,[j] collated and amplified in the present book, forms what can only be described as an unrelenting diatribe against an 'unfortunate individual'. But it is more important for us, and more characteristic of Hettner, that he did not use his chance to try to foist on to his colleagues any alternative scheme for the study of landforms; and it was most certainly not Hettner's aim, as was his colleague Passarge's,[k] to reconstruct landform geography into what he thought it should be, a fact since regretted by at least one German geomorphologist. Confident that if he could dissuade his German colleagues from adopting Davis's system as a whole they would base their future work on more secure and traditional foundations, Hettner was content with bringing out the weaknesses of Davis's position. Nothing was further from Hettner's mind than that he should foster and focus around himself a rival school of landform geography. Nevertheless, his criticism was easily represented as no more than a pedantic, embittered and personal feud with a respected foreign figure, Davis, whose contributions to geomorphology and geography had gained him pre-eminence when German geography might have hoped for a more widespread influence outside German-speaking countries.

Whatever general influence Hettner has had on English-speaking geography, through teachers such as Chisholm, Fennemann and, more recently, Hartshorne, his writings in landform geography do not appear to have had any effect on English-speaking geomorphologists. His monograph on the morphology of Saxon Switzerland seems not to have been noticed at all.[1] And such of Hettner's methodological works in this area of geography as were reviewed in the English language journals were seen only for what they obviously were: acute and stimulating, if sometimes sarcastic, critiques of Davis.[m] Perhaps the fact they were noticed at all was because they were critiques and not because they were intrinsically valuable. Whatever the case, they were disregarded, and soon forgotten.

It is tempting to explain Hettner's failure to stimulate a critical response among his English-speaking colleagues as a result of Davis's personal prestige at the time Hettner wrote. That Davis's scheme took hold among English-speaking geographers and geomorphologists so rapidly and so completely was at least partly due to the vigour with which Davis argued his case, and to the skill with which he expounded and illustrated it, to say nothing of the scheme's fascinating comprehensiveness. Yet Davis's powers of persuasion alone would not, one hopes, account for his views remaining virtually unchallenged by English-speaking students for over half a century, when their influence among German-speaking geographers was so effectively stemmed thirty years earlier, despite the strong support they got from a number of respected German geographers.

A more hopeful explanation is that Davis's scheme of the *geographical* cycle, and from which he deduced his several models of landform development, was not seen by English-speaking scholars as the threat to geography and geographical geomorphology it was for at least a few German research workers. On the contrary, Davis's scheme fulfilled its purpose, and provided a majority of English-speaking geographers with what they wanted: a standardised, systematic, qualitative and, as far as possible, genetic description of landforms not, theoretically at least, as an end in itself but as the indispensable preliminary to the geographical study of other phenomena of the earth surface.

Translator's Preface xv

By imposing on landform study a distinctive procedure in keeping with the intellectual 'climate' of the time, Davis was able to distinguish the study of landforms as part of geography and so bring geography a status it did not previously have in English-speaking countries. English-speaking geographers in general could see in Davis's system a way to advance their subject from being merely an attempt to synthesise the findings of many potentially separate disciplines concerned with the earth surface and enveloping atmosphere. For geomorphologists in particular the scheme opened up the prospect of their specialist study of landforms gaining the autonomy and prestige certain other earth sciences had already achieved, whether or not it did so within institutionalised geography.

It should not surprise us, therefore, that Hettner's criticism of Davis stimulated almost no reaction outside Germany apart from that of Davis himself and his closest colleagues.[n] Davis's scheme was so closely linked with the establishment and standing of geography and geomorphology in the English-speaking world that any criticism of it was likely to be seen as merely the spiteful reply of a German to the success of a rival American school. And by the time Hettner published *Die Oberflächenformen des Festlandes* many English-speaking geographers had already abandoned the study of landforms to specialist geomorphologists, unconcerned that the latter might not study landforms from every available viewpoint. Moreover, it was just such geographers who most vociferously demanded that geomorphologists should provide them with the knowledge of landforms they needed for 'geographic purposes'. This being the situation, Hettner's stand was meaningless and irrelevant for most English-speaking geographers. They were unable to set aside the disputatious way in which Hettner presented his case and judge it as a worthy attempt to maintain a quite different tradition of geographical scholarship. Practising geomorphologists were finding more than enough interest in landforms *per se* to worry themselves overmuch whether they were giving geographers what they wanted, or developing our understanding of landforms geographically as well as geologically and geophysically.

As their field has become a coherent and recognised earth science in its own right (a trend Davis promoted, however unintentionally) geomorphologists have grown less selfconscious of their role as scientists, and less concerned about the relations their field has with allied sciences. The floodtide of what Wooldridge described as 'the seemingly endless and wearisome argument' over the place of geomorphology as a field of learning has certainly ebbed, though not because geographers are clear about the relevance of their discipline to geomorphology. But geomorphologists at any rate have come to see how fruitless it is to try to confine their subject within logical boundaries. Instead, increasingly aware of the need for a well-disciplined approach to their work, geomorphologists are seeking to understand and improve the conceptual framework on which they base their study. Free from rival claims to their field by geology and geography, geomorphologists are more inclined to discuss the techniques and approaches they can most profitably use to reach their goal—a better understanding of landforms. Paradoxically, both the possibility of doing this and the need to do it is a belated reaction to the overwhelming influence Davis had for so long.

Since it has become almost fashionable to castigate Davis and his work, it is important to remember he has long had his critics among English-speaking students. The 'American School' of geomorphology was never quite as wholeheartedly Davisian as has sometimes been claimed. But it undoubtedly closed its ranks against the work of Walther Penck, of all people, when persuaded that it posed a fundamental challenge to Davis's system, and therefore American geomorphology as a whole. And some of the better-known, more recent criticism of Davis in English does not question the truth of the idea that landforms develop in cyclic fashion, or the value of the supposedly geographical approach founded on it, as much as stress the degree to which real conditions differ from those envisaged by Davis in his 'normal' case. There is, nonetheless, a growing awareness among English-speaking (and French-speaking) scholars of the inherent limitations of Davis's scheme, and the extent to which Davis blinded others and perhaps himself to them.

As they realise more and more clearly how much their thinking has been stifled under Davis's persistent influence, English-speaking geographers and geomorphologists are reacting in at least two distinct ways. On the one hand, students interested primarily in landforms *per se* are inclined to follow Leighly's call for a return to a free-ranging study of 'the operations of the laws of nature on the earth'.º Such students seek to free geomorphology from the theoretical limitations Davis imposed upon himself and his colleagues during the 1890's. And

Translator's Preface

their way to achieve this is to revert to a physical geography that prevailed before Davis reformed it, one in which the study of natural phenomena becomes geographical, if not a disciplined and identifiable branch of science, simply by trying to unify a number of disparate earth sciences. If, as this view implies, physical geography is no more than the attempt to assemble, in some purposeful way, the findings of one or more of the earth sciences, then clearly, the increasing specialisation and autonomy of one of these sciences, geomorphology, does not threaten geography.

On the other hand, for students concerned primarily with the human features of the earth surface, the dismantling of Davis's system is only of passing interest. They may join in the chorus of criticism, and regret the master's excessive generalisation or his use of too few, and then highly simplified, models of landform development; but they do not see his standing as a geographer undermined if we find his approach to landform study inadequate. For some such students this is because they see the essentially geographical feature of Davis's procedure in his using the interrelationships between the form-elements of a terrain to trace its development, interrelationships he claimed were expressed in a distinctive and recognisable pattern at each stage of a land's physiographic history.[p] For others it is because however erroneous and inadequate Davis's system of landform study may prove to be, this does not invalidate his view that 'the whole content of geography is the study of the relation of the earth and its inhabitants'. Geographers subscribing to this view may complain that specialist geomorphologists do not give them the kind of knowledge about landforms they need for geographic purposes. But they do not see the growing autonomy of English-speaking geomorphology as a threat to geography any more than do geomorphologists themselves. To them the notion of a landform geography, a disciplined and distinctive approach to the study of landforms, is meaningless.

But in Hettner's view the successful specialised study of any major category of phenomena, in this case landforms, is necessarily the concern of all geographers. This is not because of any secondary use to which the products of such specialised study can be put; nor is it because the study of landforms for a particular purpose makes such study *geographical*; it is not even because Hettner would want all geographers to continue to take an interest in what has long been an integral part of their field; it is because landforms, like any other class of phenomena, human or non-human, should be studied in every available and appropriate way, not least geographically.

Clearminded as to the essential character of landform geography, Hettner was sure of the distinct, if limited, contribution it could make to geomorphology. In his view the study of landforms for their own sake, and in every way open to us, was scientifically desirable; and if such study was pursued in a particular way it would, in Hettner's eyes, be geographical, even though it added nothing to our understanding of the mutual relations between Man and his natural environment. Hettner could, therefore, confidently expect geomorphology to become progressively more autonomous as an earth science unless, in concentrating its attention too exclusively on the processes that shape the land surface, it was absorbed into an expanding geophysics. He would have been able to join with Leighly in the latter's criticism of Davis for limiting the specialised study of landforms by prescriptions as to what its aims and procedures should be. But it was certainly not Hettner's view, as it was Leighly's, that physical geography should return to the state it was in during the 1870's in Germany, and somewhat later in the United States, when it was neither more nor less than a general science of the earth trying to incorporate into itself its many kindred but already autonomous earth sciences. Hettner's prolonged criticism of Davis was certainly not an attempt to undo what Davis had done for English-speaking geography in rescuing it from this state. His aim was to conserve the particular way in which Richthofen above all had lifted German geography from the same state, by applying to the study of landforms the conceptual approach inherent in the work of both Ritter and Humboldt.

The text I have translated is that of the second edition of *Die Oberflächenformen des Festlandes*, issued in 1928. The author's original footnotes are marked numerically and listed at the end of the work. Since many of these consist of almost casual references to works familiar to Hettner's colleagues at the time he was writing, but now almost unintelligible even to modern German students, I have amplified them where this seemed necessary to allow readers to refer to them if they wish. For the same reason, where Hettner gives no specific reference but names an author, and where some explanatory comment is needed, I have added a note. These are marked in the text with an asterisk and listed after the author's notes and bibliographic references, on pp. 169.

Translator's Preface

I want to take this opportunity of thanking David Davis and Ernst Plewe for much practical help and advice.

* * *

[a] Schmitthenner, H. (1956) 'Die Entstehung der Geomorphologie als geographische Disziplin', *Petermanns Geographische Mitteilungen*, 100, 257–68.

[b] Richthofen, F. von. (1883) *'Aufgaben und Methoden der heutigen Geographie'*, Leipzig: Akad. Antrittsrede.

[c] Davis, W. M. (1888) 'Geographic Methods in Geologic Investigations', *National Geographic Magazine*, 1, 11–26.

[d] Davis, W. M. (1889) 'The Geographical Cycle', *Geographical Journal*, XIV, 481–504.

[e] Davis, W. M. (1902) 'Systematic Geography', *Proceedings*, American Philosophical Society, 41, 235–59.

[f] Penck, A. (1905) 'Die Physiographie als Physiogeographie in ihren Beziehungen zu anderen Wissenschaften', *Geographische Zeitschrift*, 11, 249–69.

[g] Hettner, A. (1905) 'Das Wesen und die Methoden der Geographie', *Geographische Zeitschrift*, 11, 545–64; 615–29; 671–86.

[h] Davis, W. M. (1902) 'Systematic Geography', *Proceedings*, American Philosophical Society, 41, p. 237.

[i] Braun, G. (1911) *Grundzüge der Physiogeographie*. Leipzig.

Davis, W. M. (1912) *Die Erklärende Beschreibung der Landformen*. Leipzig.

[j] Hettner, A. (1911) 'Die Terminologie der Oberflächenformen', *Geographische Zeitschrift*, 17, 135–44.

Hettner, A. (1912) 'Alter und Form der Täler', *Geographische Zeitschrift*, 18, 665–82.

Hettner, A. (1913) 'Rumpflächen und Pseudorumpflächen', *Geographische Zeitschrift*, 19, 185–202.

Hettner, A. (1913) 'Die Entstehung des Talnetzes', *Geographisches Zeitschrift*, 19, 153–61.

Hettner, A. (1913) 'Die Abhangigkeit der Form der Landoberfläche vom inneren Bau', *Geographische Zeitschrift*, 19, 435–45.

Hettner, A. (1914) 'Die Entwicklung der Landoberfläche', *Geographische Zeitschrift*, 20, 129–45.

Hettner, A. (1914) 'Die Vorgänge der Umlagerung an der Erdoberfläche und die morphologische Korrelation', *Geographische Zeitschrift*, 20, 185–97.

Hettner, A. (1919) 'Die morphologische Forschung', *Geographische Zeitschrift*, 25, 341–52.

Hettner, A. (1930) 'Die morphologische Darstellung', *Geographische Zeitschrift*, 26, 131–6.

Hettner, A. (1921) 'Die Davische Lehre in der Morphologie des Festlandes', *Geographische Anzeiger*, 22, 1-6.

Hettner, A. (1923) 'Methodische Zeit—und Streitfragen', *Geographische Zeitschrift*, 29, 37-59.

Hettner, A. (1924) 'Noch einmal die leidigen Fastebene', *Geographische Zeitschrift*, 30, 286-90.

[k] Passarge, S. (1912) 'Physiologische Morphologie', *Mitteilungen, Geographisches Gesellschaft*, Hamburg, 26, 133-7.

[l] Hettner, A. (1887) 'Gebirgsbau und Oberflächen gestaltung der Sächsischen Schweiz', *Forschungen zur deutschen Landes—und Valkskunde*, 11 (4).

Davis, W. M. (1924) 'The Explanatory Description of Landforms', *Geographical Review*, 13, 318-21.

Lake, P. (1923) Hettner's 'Die Oberflächenformen des Festlandes', *Geographical Journal*, LXI, 461-2.

[n] Davis, W. M. (1914) 'Der Valdarno: eine Darstellungstudie', *Zeitschrift, Gesellschaft für Erdkunde zu Berlin*, 585-620; 665-97.

Davis, W. M. (1915) 'The Principles of Geographical Description', *Annals, Association of American Geographers*, V, 61-105.

Davis, W. M. (1919) 'Passarge's Principles of Landscape Description', *Geographical Review*, 8, 266-73.

Davis, W. M. (1924) 'The Explanatory Description of Landforms', *Festschrift für J. Cvijic*, Belgrade, 287-336.

[o] Leighly, J. (1955) 'What has happened to Physical Geography', *Annals, Association of American Geographers*, XLV. 309-18.

[p] Dylik, J. (1953) 'Caractères du developpement de la Géomorphologie Moderne', *Bulletin, Societé des Sciences et des Lettres, Lodz*, IV. 1-40.

Baulig, H. (1950) 'William Morris Davis: Master of Method', *Annals, Association of American Geographers*, 40, 188-95.

Author's Foreword to the First Edition

In writing this book I do not intend to draw up a systematic statement in the nature of a textbook or handbook, or provide an introduction for beginners; my aim is to review critically studies concerned with the surface features of the land, the studies themselves and their results. Such a critical review is needed when a science's methodology and viewpoint become as disunited as they now have in geomorphology (*Lehre von den Oberflächenformen des Landes*),* and there is the risk they will develop along the wrong lines. The approach and method we have received from America has been enthusiastically adhered to by many research workers, and has been accepted by a still wider circle. But despite many individual merits it seems to me to be wrong in its theory as a whole and its method in particular. For this reason I have already argued against it in a series of articles published in the *Geographische Zeitschrift* from 1911 onwards, and in doing so I enquired into other fundamental issues in geomorphology. The present book brings together not only these articles, much revised, but just as many new ones to form a critical examination of the whole discipline of geomorphology. It is intended to be a book in which geomorphology takes stock of itself; it will be appreciated that in writing it, I readily build upon personal experience. I do not claim that I have always arrived at the truth; but I hope to stimulate reflection and promote the recovery of our science which has already begun. Moreover, I want to restore geomorphology (*Morphologie*) to a close association with regional geography (*Länderkunde*), from which it has quite markedly separated itself.

ALFRED HETTNER

Heidelberg, October 1920

Author's Foreword to the Second Edition

Happily, it has become necessary to publish a new edition of this book which first appeared seven years ago. This shows that it has fulfilled its purpose in as much as it has made geographers and geologists examine critically the fundamental concepts of geomorphology. But I believe it has still not quite accomplished its aim. If it is true that the floodtide of 'modern geomorphology' has ebbed, many wrong or premature interpretations have been left behind. Anyone who intends to concern himself with geomorphological problems will still have to first become clear on basic issues of method and theory.

I had to expect that Davis and his pupils would criticise me strongly, and I had hoped to benefit from such criticism.* But I have been bitterly disappointed. He pours his cup of anger over me, and almost gives one the impression he had written comments in his copy in red ink; but he says nothing really important. It has evidently escaped him that my book does not aim to set up new theories, but to appraise critically the present state of geomorphology.

The character of this new edition remains on the whole the same. But I have made many detailed changes. Apart from correcting mistakes I have been able to incorporate the results of much new research, although I have also had to argue against new hypotheses and older opinions rehashed. In this as in the first edition I have not tried to be exhaustive. I have omitted the chapter on highlands and lowlands as being of no fundamental importance. I have greatly shortened the appendix on methods of research and presentation because it was repetitive, and because I included much of what was said there in my book on geography and geographical method.*

Professor Heinrich Schmitthenner and my pupil Albert Kolb have kindly helped me to correct the work; I am also indebted to them for compiling the index.

ALFRED HETTNER

Heidelberg, September 1928

Introduction

THE IMPORTANCE OF GEOMORPHOLOGY

The earth's surface features have long been studied by geographers, for one cannot imagine a land without its mountains and valleys, its highlands and lowlands. Certainly for centuries they were treated very inadequately and superficially; only during the eighteenth century did the development of surveying allow a correct and full description of landforms; not until the nineteenth century did the development of geology lay the foundations for a more truly scientific way of thinking, one that considered the causes and nature of phenomena. Among the greatest tasks, perhaps *the* greatest task, geography had to face in the closing decades of last century was the creation of a discipline (geomorphology) to study the earth's land surface. An unbiased judge cannot deny that much has been achieved in these fifty years. Of course there were already many accounts of landforms in the literature of geology and of physical geography, and in scientific itineraries; but anyone who compares an old geographical presentation with a contemporary one will find it hard to imagine they could be more different, both in their treatment of the earth's surface features and the extent of this treatment.

But there was another side to this development. Geomorphology was fostered at the expense of other branches of geography to such an extent that one can speak of a hypertrophy of geomorphology within geography. Like the prodigal son who on his return becomes his parents' darling, the one they cherish the most, a long-neglected aspect of a science receives exaggerated attention. For a long time almost all scientific-geographical work was directed towards geomorphology. Every young geographer had to be first and foremost a geomorphologist; the geologist was regarded as almost a born geographer. Educationalists and geologists alike wanted to link the teaching of geology and geography and to combine them in the examination syllabus. For some time now a reaction against this exaggerated view has made itself felt, and has gained ground especially under the pressure of World War I. The earth's land surface is only one of the objects of geographical study; alongside it stand at least five others of equal value, concerned with the other natural realms of reality, to which the geographer must pay just as much attention. They all need to be better understood by scientific enquiry; and from a general cultural, as much as from a national viewpoint, Man himself will always attract the greatest interest. A swing in the opposite direction was therefore justified. But it must not go too far; the baby should not be thrown out with the bath-water. The earth's surface features are at once the landscape's most prominent visual characteristics, the foundation of climate and all life, of settlement and communications, of man's economic existence, and of his cultural conditions as a whole. A geographer who does not have a clear and deep understanding of landforms will also fail to realise other geographical aims; and studies in human geography can become somewhat nebulous when their geomorphological foundation is insecure. No geographer can dispense with a thorough study of landforms; and what is true of the individual scientist is also true of the science as a whole. Geography must not become geomorphology; yet it will atrophy if it renounces it. Nor can geography content itself with what geology provides. It must build its own foundations if the whole edifice is not to collapse.

Moreover, the attitude of geologists towards study of the earth's surface features has changed in the course of time, and is still argued about. Geology cannot ignore surface features; indeed it can, as we are realising more and more, often draw from them significant inferences about internal build.* The study of rocks and processes leads almost of itself to the study of landforms. One cannot blame a geologist who, in studying an area, ends his work by considering landforms. The geographer should not be jealous of him; on the contrary, he should welcome his aid. The geographer should only demand that the geologist has the scientific training needed to solve geomorphological problems, by no means always the case today. But he should defend himself against the arrogance of some geologists who for their part, with a certain professional jealousy, want to crowd out the geographer from his field. Yet German geology has in the past contributed little to the field of geomorphology. Someone taking up geomorphological problems in the 1870's and 1880's could get little help in understanding landforms from geological literature. Even dynamic geology led a pretty miserable existence, at least as

Introduction xxv

far as exogenic processes were concerned. Even today the earth's surface features are only an extra for geologists. Geology is by its nature earth-history; therefore geomorphology is important to geology not because it studies the present form of the earth's surface, but because it enables us to know what the form of the earth's surface was like in past geological periods. This is an important but as yet little-studied aspect of so-called paleogeography, a field which will become an increasingly intrinsic part of geology. Geologists must work in this field far more than they have up till now. And since the only secure starting point from which to investigate the past is always a thorough study of the present (according to the well-known research principle introduced into science by Lyell* in particular), the geologist must be as familiar with present-day landforms just as much as with present-day processes. M. Neumayer and Joh. Walther in particular* have opened the way for such a geologically-oriented geomorphology, and in his *Morphological Analysis of Land Forms*, Walther Penck[1] has consciously put geomorphology still more at the service of geology, especially of tectonics.* He sought to determine movements of the earth's crust from landform assemblages (*Formenschatze*), an approach not normally found in the treatment of tectonics. This is the real importance of his book; it is only indirectly of service to geography.

Leaving aside its geographical (or chorological) and geological (or geochronological) importance, we may ask whether geomorphology of the earth's surface is important in its own right as a science of landforms as such, without reference to their spatial or chronological relationships. We may ask whether geomorphology exists or can exist as a separate science. The question seems proper when for the sake of comparison one thinks of pedology, a sister-science of geomorphology; while it is still included within geology and geography, pedology has undoubtedly become an independent science promoted particularly in the interests of agricultural science and forestry. Geomorphology is similarly placed in relation to strategic and transport studies. Most of our maps are ordnance maps drawn by the military, and chiefly for military purposes; the fact that military science, or at least German military science, has as yet done little to promote theoretical topography (*Geländelehre*) is a sin of omission; French military science has done more in this respect and given us two excellent books on the subject. And in the economic world we also need a technique of land appraisal (*Geländetechnik*) necessarily based on the scientific study of landforms. Thus geomorphology will certainly become more important in its own right as an object science, as a science of landforms as such. In doing so, it will treat the spatial and historical distribution of landforms as the attributes of individual landform classes, not as the object of intrinsic interest, in much the same way as botany and zoology treat the distribution of plants and animals.

It could be said that such a systematic geomorphology already exists, and has done so unintentionally for a long time. Many, indeed most, descriptions of landscape morphology by geographers have a systematic character; they give priority to the landforms themselves rather than to where these occur, and arrange their material in landform classes. Geomorphology of this kind, like systematic pedology, is a discipline within the earth sciences; but this does not make it geography. As far as geography is concerned, landforms are always a constituent part of the landscape and must be understood as such. Individual landforms have to be set in a landscape and seen as the product of a land's build and climate, and as the foundation for plant and animal life, and Man himself. The geographical approach must always on that account be chorological; it is impossible to stress this too often since specialists so often forget it. In their research work, geographers, geologists and systematic geomorphologists go a long way together only to separate at a later stage; they hew the same building stones but erect different buildings. The geographer misses his proper aim when, as so often happens, he works as a systematic geomorphologist instead of studying the earth's surface features as part of the landscape, both as cause and effect.

GEOMORPHOLOGY AND ITS MEDIA

It falls to the map, supplemented by the profile, to give a complete representation of the land's surface form. It is, after field work, the most important foundation for the geographical study of landforms. Maps have been drawn since ancient times; but only since the second half of the eighteenth century have we had both the information, in numerous measurements of altitude, and the means, with contour lines and hachures, to portray terrain. Of course there were relief models, and for the unskilled person these give the best impression;

but on practical grounds they are only of limited use. Since then, the material available has increased enormously, methods of representation have been improved continuously, and large-scale maps are available at least for developed lands. Moreover, techniques of reproduction have become so much better and cheaper that maps are widely distributed and obtainable by anyone interested in them. But we must not over-estimate their value, nor expect too much of them, nor trust them blindly. Minor landforms defy cartographic representation and even major ones* have to be generalised when the scale of the map is small. But every generalisation is subjective, and governed by personal opinions. I can recall for how long every watershed was drawn as a mountain ridge; the mountain knots of the Andes, which still haunt many a book, were no more than the outcome of such a false representation. Travellers trained as geologists have incorporated their geological conceptions into maps; these maps cannot now be used to prove the reality of such conceptions. More recent American topography is inspired by Davis's interpretation of nature and cannot therefore be used to support his views. These are the weaknesses of maps; but it is wrong to argue from this that the cartographer should work in a purely mechanical way, for experience shows that it is then that really unnatural landforms do appear on the map. Modern heads of map surveys justifiably demand that their topographers should be instructed in geomorphology so that they can correctly interpret and represent Nature's characteristic forms of terrain.

Pictorial views, whether paintings, drawings or photographs, serve to complement maps and profiles. They forgo geometrical accuracy but give instead a graphical picture of the landscape (*see* the Appendix).

As maps and illustrations have made headway, the importance of verbal presentation has changed. Originally, we had to use words to describe accurately the configuration of a landscape; today words are only a supplement to the map. What Lessing said in the *Laocoon* of the relationship of poetry to pictorial art is true of that between the verbal description of a land and a map. As word follows upon word, sentence upon sentence, it takes some time for verbal descriptions to stimulate an idea; language can never convey graphically a complex areal view. It is instructive to try to describe comprehensively a scene from a look-out point; it never succeeds, never brings the view back to mind, never gives others the right impression. It is proper that laborious descriptions of a land's surface form have justifiably become rare, and persist only in gazetteers and the like. As long as language remains purely descriptive its aim must be to bring out essentials, supplement the map or picture, and in particular to characterise the physiognomy given to the landscape by its minor landforms.

Karl Ritter and his school did not go beyond a descriptive geographical interpretation of landforms, not so much because of fundamental objections to the causal approach, as because this approach was at first impossible; and later, when most geographers were trained as historians, it was not in their line. Such an approach was not introduced into geography until Peschel wrote his *Neue Probleme der Vergleichenden Erdkunde*.* It prevailed only in the face of strong opposition from older geographers. Even now many of them disclaim the study of causes, and they are to some degree justified in doing so by the many extreme views, especially those of the American school. But the causal interpretation of the earth's surface is now an established feature of geomorphology; W. M. Davis displays a remarkable misjudgement of the situation when he reproaches German geography for contenting itself with description.* No science can do this and leave explanation to another science. Surely a full understanding of the facts is impossible without a knowledge of causes. A person who knows that a particular upland is a plateau, dissected by rivers, will understand it more precisely, commit it better to memory, and represent it more clearly to others, than someone who gives no thought to origins and takes valleys and mountains for granted. Only through the causal approach can we gain a coherent understanding of landforms and see them as constituent parts of the landscape. For landforms depend on the variety of drainage, climate, plant cover, even on the kind of animal and human life. If the causal interpretation of lands and landscapes is the proper aim of geography at all, it cannot stop short of such an interpretation of the shape of the earth's land surface.

Throughout this book my aim is to study the causes of landforms. We must, however, make one distinction at the outset. Experience has taught us that the earth's surface shape results from subterranean forces acting on the internal structure of its lands' crust (*festen Erdrinde*).* If we are to explain the surface shape of the earth's lands we therefore have two tasks. Firstly, we must interpret and explain internal build, using this phrase in the wider sense to include every aspect of the shaping of the earth's land surface by endogenic processes, and secondly, we must show how this surface has been modified; the former is the concern of tectonics, the latter of geomorphology. It is important both practically and theoretically to differentiate between these two tasks. Certainly geographers must be familiar with the facts of internal build since they are the indispensable foundation; but their investigation is more the concern of geology. The geographer should do this only when, in less well-known lands, we still lack basic geological knowledge: then he must, for a limited time, play the role of the geologist as well. Geomorphology in the stricter sense, however, not only produces results of geographical significance; it is also for preference the geographer's field of work, even though geologists must make themselves more conversant with it than they have in the past. In this book, I am concerned with geomorphology alone; I will consider tectonics only inasmuch as it will clarify the subject.

During the past half-century geomorphology has gone through a rapid, perhaps over-hasty, development. At the end of the 1860's it was still in its infancy. Certainly, English geologists had gained many valuable insights; but on the continent geology was really only beginning to study minor landforms. Peschel's *Neue Probleme der Vergleichenden Erdkunde* opened the way, but we cannot deny it suffers from a certain superficiality. Not until Rütimeyer, the Basle anatomist, published his book on valley and lake formation in the Swiss Jura and the Alps* did observation, and in particular the view that valleys were formed by erosion, become the basis for all further research; and this was necessary before surface denudation (*Abtragung über die Fläche*)* could be understood. Studies in the Cordillera of the United States were especially important, for here a number of outstanding men like Powell, Gilbert, Dutton and others worked under particularly favourable conditions of dry climate and bare rock surfaces, and were able to draw far-reaching conclusions. They were a great stimulus for Richthofen, whose study of Central Asia and China was epoch-making. His *Fuhrer für Forschungsreisende,** though now outdated in many respects, stimulated research and pointed the way more than any other work; it is still today the best introduction to the study of geomorphology. Only someone who has consciously experienced the earlier period can measure the immense progress of geomorphological science in recent decades. The period of accurate fieldwork by individuals and the growing development of methods had begun. Neither the comparative study of maps nor the deductive approach based upon the nature of processes can be discarded; but they are never used except in conjunction with field observation. As research has proceeded, we have come to recognise more and more the immense importance of the earth's surface processes; they do not merely shape the details of this surface, they create an entirely new one. Yet at the same time we have not lost sight of the fact that internal build conditions the work of these processes, and that only by studying this too, can we understand landforms.[2]

THE DOCTRINES AND METHODS OF W. M. DAVIS

Of course, opinions differ widely on particular issues, such as glacial erosion, the role of wind in deserts, and marine abrasion or sub-aerial degradation as the cause of planation. But the first fundamental split, involving the very nature of the science, has come about through the teaching of W. M. Davis.[3] He took as his starting point the results of research in the Cordillera of the American West and then gathered a rich harvest of geomorphological experience on many journeys all over the world. Only later did he begin to work deductively; but deduction is now an essential and prominent feature of his work; use of the deductive method, or explanatory description, has become the methodological idiosyncrasy of the 'modern school'. Another theoretical concept goes hand in hand with it, and the connection is not fortuitous. The decisive theoretical idea is that the land is progressively degraded and levelled by forces working upon its surface, and that this levelling is interrupted by new uplifts. Although this idea is not in itself new, it is now advocated with much greater perseverance and partiality. The concept of morphological age, the cycle and the peneplain (*Fastebene*) to which it gives rise, stand at the centre and are the catchwords of Davis's school. Internal build and climate are relegated, as the bureaucrats would say, to a subordinate place among factors influencing surface modification, with the result that the truly geographical viewpoint is virtually suppressed.

As a result of his great didactic skill, supported by a rare ability

Introduction xxxi

to draw, and the alluring simplicity of his doctrine, Davis's views have been as widely accepted as was Hegel's philosophy in its day. Certainly this has varied from country to country. American research has long proceeded wholly under the banner of Davis's method and doctrine; the sober English have largely discarded it. On the other hand, many French research workers, in particular de Lapparent[4] and de Martonne,[5] and a large proportion of German geographers and geologists, have more or less capitulated to it. Thus today, or at least until recently, when there has been a certain amount of clarification, there are two parallel schools of thought in geomorphology. We have to decide between them and recognise the path which geomorphological research should follow.

THE AIM OF THE BOOK

This book is not intended to be a systematic geomorphology: it discusses only the fundamental issues. To make a positive contribution one could write a history of geomorphology, for which there is, as yet, little material to hand. It would be very instructive to see how the scientific approach to landforms has gradually developed; the history of science does not deserve the scorn so often doled out to it, for it sharpens the understanding, suggests new problems, warns of erroneous ways, and justifies those earlier researchers who expressed much of what is today thought to be new. It would be of the greatest interest to see how earlier workers tackled such particular problems as the development of karst, the surface configuration of deserts, glacial erosion, or the formation of remnant surfaces (*Rumpfflächen*),* as well as to recognise the characteristics of particular schools, for example the way English geologists handled geomorphological problems, or the Americans those of the Cordilleras.

Here I have chosen another way. Without maintaining a strictly historical approach, but following in some fashion the development of geomorphological thought, I will methodically consider the subject's most important problems in turn, and investigate to what extent they have been solved and what remains to be done. I begin with the simpler problems and go on to the more difficult and complicated ones, finally arriving at a complete appraisal. Such a treatment could be entirely academic, platonic, objective, in the limited sense of the word; it could compare various opinions on the questions at issue without taking sides. However, I do not believe that this would render much service to geomorphology at this time. I shall not simply report, but rather take up a critical position.[6] If the confusion in geomorphology is not to become still greater, we must examine the value of methods and theories. We must oppose anything that furthers false methods and factual errors so that the ground may again be clear for sound scientific research. I hope that such a study will benefit geography as well as geology and systematic geomorphology.[7]

CHAPTER I
The Minor Features of the Landscape

THE IMPORTANCE OF MINOR LANDFORMS

The layman's attention is first attracted by peculiar minor landforms; he finds them remarkable and wants to investigate them, while he takes major landforms and the common form of slopes almost for granted. Geomorphology developed in a similar way. It began by studying impressive cliffs, limestone caverns, and sandstone and granite stacks. It was oriented towards these minor landforms rather than towards such major ones as valleys, which were long considered no more than fissures; the study of minor landforms seemed an accepted part of the science. But recently, under the leadership of W. M. Davis, many geomorphologists have ceased to study them. Not only are such impressive rock forms as, for example, earth-pyramids, caverns (*Grotten*), natural arches (*Tore*), rock pillars (*Pfeiler*) of the Quadersandstein,* or the boulder fields (*Felsenmeere*) of other uplands not given a word of consideration, but neither are the landforms of different rocks and climates discussed in any comprehensive way. They are not only neglected; they are as completely disregarded in the detailed studies as in the summary writings of this school. Rühl explicitly declares that it is unnecessary to study them, relegating them to the area of geology.[8] And yet they are important in two respects: directly, because they are a determinative element of the landscape, indirectly because they often provide the key to understanding major landforms. For this reason the older German school of geomorphologists under Richthofen's leadership always paid great attention to minor landforms. I myself began by studying the distinctive rock forms of Saxon Switzerland, Walther studied the minor landforms of deserts, and their study is a central theme of Passarge's physiological geomorphology, in which he indeed over-emphasised them at the expense of valley formation. Thus even at this level, the divergence of approach, now a feature of geomorphology, is evident.

The Surface Features of the Land

To a considerable degree minor landforms determine the appearance of the landscape, its physiognomy, its landform assemblage (*Formenschatz*), its structural style (*Baustil*) and thus even its aesthetic character. How can we imagine Saxon Switzerland, the Harz or the Riesengebirge without thinking of their typical rock forms? How can we imagine high mountains or deserts without theirs? Many areas lack impressive minor landforms, but then it is precisely the absence of these which is characteristic; and why smooth slopes should develop instead calls for detailed study. Without minor landforms a geomorphological description is dead and schematic; it is equivalent to the small-scale map, which is unable to show the distinctiveness of minor landforms in the way a large-scale map would. What we feel at first to be a disagreeable necessity is soon willingly abandoned without justification. Moreover, the influence of landscape upon the plant and animal world and on Mankind, in other words much of its geographical importance, is thereby to a large extent lost.

Furthermore, minor landforms have a second, indirect importance. By helping us to understand the processes which modify the earth's surface they help us to understand major landforms also. They are brought about by the weathering and mass-wasting (or mass-movement) which we can therefore recognise by studying minor landforms as well as by studying the nature of the regolith (*Bodenbeschaffenheit*).* For example, the importance of percolating water in Saxon Switzerland can be recognised by studying the small caverns. Most minor landforms are relatively young and short-lived; but although they disappear, the vestiges they leave behind accumulate to form major landforms. Valley sides are the result of the cumulative effect of the processes which engender minor landforms; thus their form is related to minor landforms. The free rock faces of porous sandstone are a good example of this. To understand large-scale erosion and major landforms, we must study minor landforms as well as observe the processes of weathering and denudation. We must use a physiological approach, to use Passarge's phrase. The extreme one-sidedness of Davis's method, its schematic nature, and the way it disparages the diversity of landforms in different parts of the earth are, in the final analysis, the result of his neglect of minor landforms. It is wrong to say that these processes are well enough understood for us to be able to take them for granted. On the contrary, not until quite recently have we made any real progress in understanding them, and much still remains to be done.

METHODS OF DESCRIPTION AND INVESTIGATION

As in any scientific study the first requirement in geomorphology is the exact description of what is seen or, in any way, sensed. Description is in some measure objective and lasting, while explanation is subjective and liable to change. Certainly, it is by no means easy to describe minor landforms. General words like flat, level, steep, rocky, etc., are a help, as are comparisons with everyday objects; but they are makeshifts and do not allow us to formulate general laws. Even maps fail where minor landforms are concerned. It is here that we are most justified in using illustration and especially photography. Chaix's collections of geomorphological photographs pay special attention to minor landforms;* one can scarcely go far enough in trying to reproduce characteristic minor landforms, even though this is often very difficult to do.

Explanation, in other words tracing phenomena back to their causes, should begin only after the phenomena have been described. Explanation must not be introduced into description, because then it is easy for description to be falsified. To explain the features of the earth's surface we must trace them back to the processes to which they owe their formation. Since the processes involved are largely physical and chemical ones our studies must fall back upon physical and chemical laws, and the day will come when processes can be treated as part of applied physics and chemistry. Dynamic geology must study these processes since, according to the principle of actualism, only through these will we be able to understand the geological past. Moreover, geography must do the same, for only in this way can minor landforms be causally related to the character of the landscape as a whole. But geography must not remain a science of processes. Few landforms can be traced back to a single process; most result from the combined effect of several processes of weathering and mass-movement. Geomorphology is something other than dynamic geology.

The origin of a landform can be most readily and directly observed when the process is very intense, *e.g.* a landslide, a dust-storm or a flood. Observations of these processes are of great value and reports of them in newspapers or local archives should be collected whenever possible (E. V. Hoff started archives over a century ago.)* But through long, painstaking observation, like that with which the Müller brothers studied fertilisation by insects, or Lubbock and others the life of ants, we can discern even minor processes and their effects: stone fracturing in deserts, the wasting of moss cushions by being covered with sand, sheet-wash and gully erosion after downpours. Walther, Passarge, Sapper and many others have advanced geomorphology considerably by such observations.

Generally, however, we have to approach the problem indirectly. If we establish the distribution of the landforms under study in a particular area and, within limits, their distribution beyond this area over the entire earth, and go on to compare this distribution with that of other phenomena with which we can presume the landforms are causally related or with the conditions in which they are found—for example, location in sunlight or shade, to windward or leeward, or association with mist, groundwater or spring lines—we can gain important positive or negative evidence as to whether two phenomena are either causally related or, in fact, unrelated. But we must be cautious. When we begin our study of distribution our understanding not only of the landforms themselves but also of the causal processes involved is often incomplete; as a result we find both partial correlation and partial deviation in the distribution. Only our efforts to explain this inconsistency will lead us to a clearer understanding. When a phenomenon occurs in an area without being associated with its usual causal process, it is not compelling evidence that the one does not depend on the other; the landform may date from an earlier period and climate when the process was effective. Passarge and others actually regard this as the rule rather than the exception.

The comparison of surface features with the nature of the regolith can be an invaluable aid. Surface form is not a result of the regolith, nor is the reverse true; but both depend on the same processes. If, therefore, we know or establish the processes which formed the soil, we can also more readily appraise the surface features associated with it. Débris *in situ* or transported, river deposits, boulder-clay, salt efflorescences, calcretes, patterned ground (*steingärten*) in polar lands and so on, and the characteristic landforms which accompany them, may serve as examples. Geomorphological studies can never be completely separated from the study of the regolith.

If comparative study, or even merely conjecture, leads us to think we can take it for granted that a landform is caused by a particular process, we must demonstrate the way the process works. Walther, for example, and the research workers influenced by him, did not do this adequately enough when they preached the gospel of wind erosion. On the one hand, they found particular landforms in desert areas and, on the other, noted that in these areas the wind blows effectively. So they ascribed all desert landforms to wind. But how can ramified caverns penetrating deep into the rock, or tortuously winding valleys, be deflated by wind? Forces other than wind certainly operate in deserts, and if necessary we must invoke past conditions in order to explain the present.

CLASSIFICATION AND TERMINOLOGY OF MINOR LANDFORMS

By describing and studying them we can arrive at a classification as well as a terminology for minor landforms. In the first instance, this will be purely descriptive, based on external characteristics of form as well as material composition. But as study proceeds, the classification of landforms will increasingly take account of origin and thus become genetical. Such a classification will greatly facilitate description of the landscape's physiognomy, and it will greatly assist the comparison of different landscapes. Every attempt at classification must keep this in view if it is not to miss its purpose and become mere pedantry. It must keep to the fore properties that are geographically characteristic and effective; it must at the same time bring out what is common to landforms which together comprise a landform assemblage. While a purely descriptive classification may at first distinguish, say, positive and negative landforms, and then subdivide the latter into horizontal and vertical depressions, or separate those open on one side from those open on several, and so on, this is to be seen as no more than a schema, one that must be filled out by a more penetrating discrimination which takes account of causes.

THE ACTION OF FORCES AND THE ANALYSIS OF LANDFORMS

To begin with, all curious landforms—and only these were considered —were seen as freaks of nature or were explained as the result of catastrophic events: volcanic eruptions, great floods and the like. That is the way of science, and every traveller who talks about striking landforms with neophytes of landscape, or even with educated people, will come up against the same attitude. To attribute all inland cliff features (*Felsgebilde*) to the work of the sea, when it was more extensive than now it is, was a step forward from this. Even Lyell, the great advocate of actualism who always sought to base his work on present-day processes, made the mistake of unhesitatingly accepting the view that the sea had also affected inland areas, despite the fact that Hutton and Playfair had already gone beyond this one-sided approach. Only gradually, and not until the second half of last century, was the view that the sub-aerial forces of inland regions are also effective in shaping landforms generally accepted.

Here we are not concerned to set out the way in which particular forces work, but only with the fundamental idea: to distinguish the several processes and their interaction. The main point is our knowledge that processes are multifarious, and that almost always several of them work together to shape a landform. Only exceptionally, therefore, can a landform be considered the product of a single process. It may be correct for dynamic geology to focus its study on processes and put landforms in second place; it is inexpedient for geomorphology to do this, for it does not give us a trustworthy knowledge of landforms. On the contrary, geomorphology must start from the overall nature (*Gesamtheit*) of processes interacting on a particular part of the earth's surface, and explain the area's landforms from this. The landforms themselves and the landform assemblage of an area are the phenomena studied in geomorphology. Geomorphology does not, for example, have to consider the different landforms produced by wind in the most varied parts of the earth—that is only an auxiliary study—but with desert landforms, whether shaped by one process or another.

We can divide the formative processes into those of weathering and mass-movement, transportation and relocation; but the two classes of process interact with one another. For movement to take place it is usually necessary for rock to have been first disintegrated by weathering; disintegration in turn initiates a new phase of weathering.

Even though the principles of weathering, or, as W. Penck would say, of 'rock reduction', have long been known in broad outline, modern pedology first taught us to understand the processes in detail;

The Minor Features of the Landscape

geomorphology must use the pedologist's results. Processes are much more diverse, more difficult to understand, and therefore more important than was originally thought. Weathering does not provide everywhere the same amount of prepared loose material and make it available to the processes of displacement.

Moreover, mass-wasting or mass-movement (W. Penck's 'transference of material') are still by no means fully understood.[9] Since studies were first carried out in the moist climate of the temperate zone, it is understandable that they began by emphasising the role of water in bringing about such movement. It is understandable too that the effects of water trickling or washing across the surface was observed first, and only later the effects of groundwater. But in precisely this field we have gained very important results in recent decades; we have come to know of so-called 'creep' (*Götzinger*),* or to use Passarge's term 'soil thrust' (*Bodenschub*), and recognise the way material can be washed out of débris by becoming saturated or by water trickling through it. The way in which sloping surfaces are gullied has, of course, been recognised for a long time; the smooth form of many slopes, a smoothness that cannot but be seen as the work of trickling water, is more difficult to understand. But we could not explain the degradation of very gentle slopes at all. Here Schmitthenner's study of dells[10]—rectilinear, quite shallow depressions without the floor a valley has, formed by surface run-off, and continually refilled by soil creep—seemed to point the way.* It has now been realised that all these processes have a much greater effect than had been suspected, and that hillsides and even upland plains (*Hochflächen*)* owe their form to them. In reality they play just as noteworthy a role as erosion by running water, and must be studied just as much.

In our climate wind has generally little effect; but it must be considered. It plays a significant role not only on the sea-coast and in mountains but on the stubble of arable fields and on roads. In dry regions wind is as important as water, although the role of the latter should not be underestimated. Opinions are still divided on the nature of wind action. If Walther stresses the way wind lifts material and carries it along, if he 'sings the song of deflation', as Passarge puts it, the latter sings that of sand corrosion or blasting. According to Erich Kaiser,[11] with whom I agree, each process has its proper place: sand blasting shapes specific minor landforms, but only through its ability to move sand and dust does the wind contribute to the formation of major landforms.

This has not exhausted the processes which move material; but since we are here concerned only with the basic approach we can leave the rest aside.

Any analysis of processes can end in a genetical classification. But I will refrain from carrying this out or even trying to do so. Our knowledge is still too fragmentary to make a satisfactory solution possible.

THE DEPENDENCE OF LANDFORMS ON ROCK-TYPE AND DISPOSITION

That landforms are related to rock type was recognised quite early (in Germany in particular by Cotta);* indeed, it was over-emphasised at the expense of other factors. Landforms were seen as a direct outcome of rock type which could not and need not be further explained. We have come to see this as too one-sided and insufficient a view, and we now recognise that landforms vary with the way rock is disposed (*lagerungsverhaltnisse*)* and with climate, that the landforms of mountains, deserts, and steppes are different from those of our own area. But any observation in the field will show that there is nevertheless a marked dependence upon rock-type. But we cannot simply take this for granted. We must explain it in terms of the rock's physical and chemical composition and its structure (*struktur*). It is not enough to distinguish between hard and soft rocks, meaning by this not actual hardness or softness as shown by a blow of the hammer, but greater or lesser resistance. Our study must go deeper than this. As a result of their structure and especially their jointing, many kinds of rock tend to break down into distinctive shapes (*Absonderungsformen*) which in turn influence landforms. The columnar jointing of basalt, the rectilinear block-jointing of many sandstones and granites, or the foliation of gneisses and crystalline schists come to mind. Chemical composition determines whether disintegration or decomposition is the predominant form of weathering, and therefore whether its outcome is talus, coarse sand or loam. The permeability of rocks and soil is especially important, since on this depends whether water flows away on the surface, remains in the soil, or sinks into the rock, not to reappear until it reaches impervious strata. Sheet-wash (*Abspülung*) predominates on impervious rocks; it gives rise to slopes more or less smoothly planed and occasionally cut by gullies, on upland plains to dells. On permeable rock sheet-wash is slight; accordingly slopes are undercut and break away to form free rock faces which are slowly displaced backwards to leave gentle talus-covered slopes at their foot (*fläche Fusshalden*). Solubility, the property which distinguishes pure limestone from most other rocks, and the degree to which a rock is subject to chemical decomposition, are also important. To understand landforms the whole complex of characteristics must always be taken into account. According to circumstances a particular type of rock gives rise to different landforms; but all are stamped with a common impress. Within any one kind of climate (*gleichen Klima*),* it is possible to speak of granite, schist, massive sandstone and so on as having a characteristic assemblage of landforms.

Since rocks of any one kind, petrologically speaking, have the same characteristics, it is generally true to say that when exposed to the same kind of processes they give rise to the same kind of landforms. But sometimes small differences of composition (*Zusammensetzung*), differences which escape petrological definition and rock grouping, can result in marked differences of landform. Sandstone gives rise to quite different landforms according to its cementing material. In the Odenwald, a resistant granite generally forms eminences, a less resistant one depressions. Even these small differences will be recognised and taken into account by petrology when it is more closely associated with the study of surface features than it has been hitherto. The more general rock-classes have little geomorphological value since differences of origin and of characteristics of geomorphological importance do not go hand in hand. How differently clay and sandstone behave, though both are classed as sedimentary rocks!

And even in any one kind of climate we must also consider not only the petrology of a rock but also the way it is disposed, the dip of the strata and so on, since on this depends what surfaces are made available to the attack of modifying forces. The geomorphologist will soon learn to perceive these differences in the field when his eye for them has first been sharpened by studies in geology. Without such observation, research lacks a proper foundation.

THE DEPENDENCE OF MINOR LANDFORMS UPON CLIMATE

We were slow to recognise that landforms are related to climate. Differences of climate in the settled areas of central and western Europe are generally too slight to draw our attention directly to the differences in landforms they cause. But the geomorphological importance of even the greatest contrasts in European climate, namely those resulting from the variation of climate with altitude, were long overlooked; the singularities of mountain landforms were attributed solely to the available relief and the steepness of slopes.

Studies in desert climates paved the way for us to understand how landforms are related to climate. It is true that Richthofen recognised the importance of wind in dry regions; but since he did not travel in true deserts, but in areas adjoining them, he considered it important because it was responsible for the dust deposits of these areas. Desert landforms as such were first studied in detail by Johannes Walther, following Schweinfurth's lead.* He was soon followed by a succession of other researchers, and he himself extended his studies to other deserts. In the Egyptian desert he re-examined, and in doing so moderated, the exaggerations of his earlier views.[12] It is to these studies that we owe the important realisation, now seen to be true of other climate regions as well, that the forms of the earth's land surface vary with climate, that each climate has a particular assemblage of landforms. But in realising this we have not quite cleared one hurdle. When it was seen that particular landforms occur in desert regions they were prematurely taken to be specifically desert landforms (*see* p. 5), and identical or similar landforms in other parts of the earth unhesitatingly spoken of as the product of an earlier desert climate. But an analysis of formative processes shows that other climates can also provide, by a kind of convergence, the conditions needed for their formation. This is the case, for example, if the rock is very porous and there is no surface run-off of rainwater, and if, as with sandstone, it is less liable to chemical decomposition than to mechanical disintegration. In much the same way as plants are affected by soils, lithological facies and climate can to a certain extent simulate one another; dry rock and dry climate produce similar landforms.

With the way opened up by the study of desert landforms, the characteristics of formative processes and landform assemblages in other climate regions, including those of the temperate zone, were examined. Passarge's investigations in the sub-humid steppe-desert, and Sapper's in humid tropical lands, have been followed by studies in the most diverse regions. But here a wide field is still open to geographers. Climatic geomorphology can be compared with study of the way soils or vegetation are related to different climates. These studies were put on a scientific basis only in recent decades. But just as they have already made great progress and established the facts, at least in outline, the physiognomy and physiology of landforms in different climates will soon be an assured part of geomorphology.[13]

It is precisely here that the need for a geographical interpretation, and for it to be considered as valuable as a geological one, is most apparent. Insofar as surface features are a result of rock type and rock disposition, the geologist studying rocks all the time has an advantage over the geographer, for whom a knowledge of rocks is only ancillary. In assessing how surface features depend upon climate the situation is reversed; this is geographers' work, not geologists'. It is not enough to collect hurriedly together some figures of mean temperatures or the magnitude of temperature ranges or rainfall. We must take account of the overall character of the climate as it is formed by the integration and interaction of warmth and indeed all climatic factors. This must be coupled with the influence of plants and animals, for they are related to climate in quite specific ways. A training in climatology is just as necessary to the study of landforms as a training in petrology.

MINOR LANDFORMS OF THE PAST

An important and yet most disputed question is to what extent we may and must reckon with a different past climate to explain minor landforms. That we must do so as far as major landforms are concerned is unquestionable. Valleys and surfaces of degradation (*abtragungsfläche*)* are of Tertiary or even greater age. Not only is the terrain (*Boden*) of the whole of northern Germany a product of the Glacial Epoch (*Eiszeit*)* but the valleys of the Alps are without doubt glacially modified. The features of a steppe climate have also been left behind. But in general, minor landforms are short-lived and the climate of past millennia has in its essentials probably been much the same as that of the present. The greater changes of climate took place much earlier. For that reason alone there is room for uncertainty on the question, and opinions are still divided.

There can be no doubt that many minor landforms of the Central European landscape have been derived from an earlier climate. The polished and scratched rocks of the Alps and Northern German Lowland, moraines, the peculiar kettle-holes in boulder clay, etc., are certainly the remnants of the Last Glaciation, but only of the Last and not of earlier glaciations. Many débris features and boulder fields may have a periglacial origin, that is to say, may have been formed by soil flow (solifluction) in the raw and snowy climate that prevailed during the Last Glaciation at the margin of the great inland ice-sheet. But why should block streams and boulder fields not originate in our present climate as well? And perhaps conditions were especially favourable for the development of tors (*Felsenbildung*) in an earlier period and many may have been preserved from this earlier period; but the arguments propounded by Obst in his study of the Heuscheuer mountains* against these formative processes continuing today seem to me to be rather unsound.

If then, the preservation of individual minor landforms from the more recent geological past is plausible, this is not to say that all or even most present-day landforms are relic landforms, and that landform-shaping processes are now quiescent, as Passarge and his followers believe. The historical evidence which they cite in support of this view is weak; on the contrary there are many reports in archives, etc., of the fresh initiation of landforms; we can directly observe how natural processes still modify stone buildings. Even in historical times, before the beginning of modern afforestation, our mountains were unwooded for decades, the soil exposed to wind and weather. Winters are now less rigorous than they were in the Glacial Epoch but rigorous enough to cause a good deal of frost weathering; and summers are warm and moist enough to favour chemical decomposition. The soil-creep of the temperate zone is, without doubt, less intense than solifluction in polar zones. But after the painstaking studies of wooded areas by Gözinger, Schmitthenner, and others, it can scarcely be questioned, despite Passarge's objections, that creep, together with soil erosion, displaces an appreciable amount of material. Pedology explains present-day soil formation as a result of present-day conditions of climate; since the origin of minor landforms always goes hand-in-hand with soil formation, they too must form under present day conditions. In attributing the striking rock forms of the *Quadersandstein* of Saxony and Silesia, or the Bunter Sandstone of the Palatinate to a former desert climate, too little account was taken of rock composition, and of the fact that a particular climate and a particular rock composition can simulate one another in their effects, that a porous rock is as dry as rock in a desert. It is inconsistent of Walther, following what he rather unfortunately termed the 'ontological method', to refuse to allow that deserts once had a moister climate and then with other researchers who follow him, to assume unhesitatingly that Germany had a desert climate even in the recent past.

One must also be on one's guard against immediately concluding that there has been an overall change in conditions where different landforms are combined. In the Algerian Tell-Atlas, and similarly in the Appenines and the Crimea, smooth hill-slopes cut by rain furrows can be observed. It might be inferred from this that the smooth slopes belong to an older period of formation, the furrows to a younger one. Of course the furrows of any one part are younger than the smooth slope into which they are incised; but if because of this a more recent period of furrow development is contrasted with an older one of smooth-slope development, too little account is taken of unperiodic and purely local changes. A surface which slopes gently as a result of surface run-off and soil creep can be lacerated into runnels and gullies by a heavy downpour of rain; then it can gradually heal over again. In the course of erosion river channels shift; where the river is currently impinging, gulch development may begin, and it stops where the river no longer flows. We would be entitled to conclude that the climate of an area had changed only if all the changes in the area's landforms had the same tendency and consistency either of fresh runnels being formed, or slopes being healed over and smoothed out, but not both.

CHAPTER II
The Origin of Valleys

THE EROSIONAL NATURE OF VALLEYS

The problem of the origin of valleys was second only to that of minor landforms in the development of geomorphological research; it was in solving this, for long the central problem, that geomorphology attained its present position. Two opposing views were held as to the origin of valleys. The eighteenth-century Neptunists propounded what was admittedly a very crude theory of erosion, one invoking great floods; but in the nineteenth century the Plutonists' theory, according to which valleys originated as open fissures during mountain uplift, and were only modified in detail by water and the forces of weathering and degradation, gained the upper hand. Although it is true that Lyell rejected the fissure theory,* he did not adopt Hutton's and Playfair's explanation of valleys as the products of flowing water; he saw them as marine features. Only gradually was the view pioneered in England (Greenwood 1857, Beete Jukes 1862), that valleys had been eroded by the rivers that occupy them in a process actively continued over thousands of years. But it was some years before this view gained ground on the Continent. As late as 1867, Peschel advocated the fissure nature of valleys in a brilliantly-written article embodied in his *Neue Probleme der Vergleichenden Erdkunde*. But this viewpoint was on the defensive and could not prevent the triumph of the erosional viewpoint. At about the same time (1869), Rütimeyer introduced the theory of erosion here in Germany, in his book on valley and lake formation.

The principal argument in favour of the fissure theory and against that of erosion was obviously the seeming disparity between the size of valleys and the size of rivers; rivers did not seem capable of shaping such large excavations. But this argument did not allow for time. Drips wear away stone. A river must use its available force to work and eat into its bed; if only we allow it a sufficiently long time it will be able to excavate a valley. And a river, in its natural state, its

The Origin of Valleys

flooding unregulated, undoubtedly cuts and removes more than one subjected to man's controlling hand. The river itself need only work at its bed; bevelling (*Abschrägung*) of valley-side slopes and the resultant widening of valleys is a matter of weathering and denudation by surface run-off and other processes of mass-wasting.

It is indeed true that valleys are often seen to follow tectonic lines: troughs, faults, rock-clefts. Some thought it possible to generalise from this fact and assume that open fissures existed where none could be seen; but then open cracks of the kind formed when a loamy surface dries out would also be difficult to observe. It is undeniable that in many instances detailed study has shown fissures or tectonic lines to be the most likely causes of valley alignment where they had not at first been expected. Many geologists, Deecke in particular,* now prefer to use this fact as an argument for discarding the theory of erosion and returning to the fissure theory. But they are making a mistake. If a fault or some other tectonic line coincides with the alignment of a valley, it probably proves, or at least suggests, that it has influenced the course the river has taken. But this is not to say that the valley was an open fissure available for the river's immediate use. In any event, if there were such a fissure it could not have been very deep; the present sinuous valley shows nothing to indicate that it was once a deep open fissure; it must have been excavated by the river. While the smaller gullies of the Elbe Sandstone Mountains are associated with rock clefts, genuine and, on the whole, markedly twisting valleys have nothing to do with them. The generally narrow valleys in which rivers, after having flowed across lowlands, break through a mountain or mountain range afford special support for the fissure theory. At first sight they are certainly difficult to explain by the theory of erosion, so much so as to cause Peschel to abandon the theory; but we shall see later how those who have understood the problem have come to terms with it in various ways.

Advocates of the erosional nature of valleys base their case firstly on the fact that in most valleys there is no sign of a fault or tectonic line of any kind. They claim, in other words, that in most cases fissures are not observed facts but merely hypothetical constructions. On the other hand, they can claim that the power of flowing water to excavate is an observed fact. After every thunderstorm we can see running water cut runnels; veritable valleys are incised into recent volcanic or other depositions after only a few years. If water can effect such work in this instance, why should it not be able to do likewise in other rocks? Where rock is harder the work is slower, but it can be achieved with time; and knowledge of the earth's history as a whole constrains us to reckon with a considerable time span. The form of many valleys manifestly points to the work of water; on seeing Alpine ravines one can scarcely question that they have been cut by water. How can we otherwise explain why valley meanders are larger or smaller according to the river's size, or the striking contrast between the undercut bank on the outer and the slip-off slope on the inner side. Were it a case of open fissures, a large river might use a narrow fissure, a small river a wide one; in fact, a large river always has a large valley, a small river a small valley, so that the size of the valley corresponds with the size and energy (*Arbeitskraft*) of the river. Large rivers usually have a gently inclined valley bottom; small ones, on the contrary, have a steep gradient. Where several open fissures adjoin they would differ in depth; the fact that valley bottoms are generally at the same level (a fact to which Heim, and rather later Playfair), drew attention as a specially noteworthy phenomenon as early as 1791, can only be explained as the work of erosion. That old valley floors are often at a considerable height above rivers and covered by river gravels is evidence that would persuade even the sceptic; to imagine that the river once flowed at this high a level and that a fissure suddenly opened beneath it is, of course, just too improbable.

Thus, there is an overwhelming amount of evidence for the erosional nature of valleys, for the view that they are negative landforms shaped by rivers themselves. This is not to say that they are unrelated to internal build, in particular to tectonic lines, as was thought to be the case for so long. In many cases their layout seems to have been determined by tectonic lines, but only their layout, and not their deepening. River alignment is influenced by tectonic lines, but rivers themselves excavate their valleys along a course allotted to them.

At this point we can formulate the law: valleys are the work of the rivers flowing within them. But not every elongated hollow in the landscape can be taken as a valley, as is often done in non-scientific circles and such features as, for example, the Upper Rhine Plain or the trench fissure of the Ghor, treated as valleys; they are trenches of tectonic origin, merely used by rivers. Moreover, they differ both in size and form from true valleys, and valleys have other characteristic properties apart from their elongated form, in particular the uniform slope of their bottom. We must select a definition; for science always has the right and obligation to use more precisely terms taken over from colloquial use. It seems to me that the word 'valley' is useful scientifically only if we use it in the narrower sense. How can we use the single word 'valley' for true valleys, trenches, and elongated troughs? Valleys in the narrower sense of the word are the work of running water, and we can therefore incorporate such an origin into the term, and so convert it from a descriptive into a genetical definition.

In most areas of the earth we find an infinitely large number of true valleys; but it is not always clear whether an elongated hollow should be called a valley or not. The Yosemite valley was long considered a trench and only later recognised as a glacial valley. It is not yet certain whether the Nile valley in Egypt is a broad valley or a trench. Many transitional forms can be found between gullies cut into a valley side, which no one would call valleys, and true valleys. The shallow 'dells', which frequently pit the surface of an upland plain, have often been called valleys and do in fact become such; but apart from their shallowness they differ from valleys in not having a valley floor. Special valley forms are found in formerly glaciated regions and in deserts.

At the beginning of this chapter I referred to the general importance of deciding between the fissure theory and the theory of erosion. The triumph of the former, and with this the denial of any major effect to running water, would have meant that the earth's land surface as a whole was virtually determined by the forces of the earth's interior and that sub-aerial forces only modify it in minutiæ and detail; geomorphology would have more or less become tectonics. Initially, the success of the theory of erosion meant no more than that valleys are formed by the force of running water acting on the surface; but it is only a short step further to the idea that the rest of the land surface has also been greatly modified by sub-aerial forces. The extent to which sub-aerial forces do modify the surface has, in fact, been recognised more and more. With the triumph of the theory of erosion, geomorphology became an independent discipline distinct from tectonics.

THE THEORY OF EROSION

Until now the study of valleys has been essentially inductive and, since experiment is rarely practicable, founded on the comparison of observed facts. This is not to say, however, that it dispenses entirely with deductive reasoning; this is scarcely the case in even the most extreme instance of induction being used. Once it had been realised that valleys were the work of running water, ideas on the way in which water had its effect were continually being devised as aids to further study. But they remained secondary, and were certainly slow to develop; they would never have done so without the constraint of knowledge gained by induction. The apparent contradiction between the scale of the effect and the smallness of the force, which indeed becomes a significant factor only because of the long period involved, was much too great for earlier research workers to be able to look at a river and conceive that it could achieve so much; it still is for the uninstructed. But this seeming contradiction made it all the more necessary to study the process exactly, and to visualise it in all its details.

Hydrodynamics, the discipline which studies the movement and carrying capacity of flowing water, must be our starting point. The link with geomorphology was not easily made, and is still perhaps somewhat incomplete; but I cannot clearly understand how W. Penck can say that geomorphological research has until now neglected physical methods.[14] Hydraulic engineers consider primarily the movement of rivers as they are now, and the shape of their beds. But in doing so they also provide the basis for understanding how valleys come about in the course of a long geological development. G. K. Gilbert, studying the American Cordillera, provided the first comprehensive theory of erosion; Richthofen, Noë and Margerie, Penck, Philippson, myself and others developed it further and gave it systematic formulation.* Davis and his pupils have taken strikingly little interest in this theory, although the whole of their deductive argument rests on the view that running water erodes; they see erosion as an established fact needing no further study. Likewise, geology has until recently neglected the theory of erosion; often geologists do not bring out clearly enough the nature of incision and the limits to the work erosion can do, with the result that they have sometimes reached wrong conclusions and conceptions which could easily have been avoided by rethinking the whole theory.

Even now the theory of erosion still contains a number of debatable tenets, and one must be wary of applying them dogmatically. They need to be more intensively compared in every particular with the facts as they really are, and examined by the inductive approach. Most young geomorphologists devote too little work to this solid foundation. We should use experiment in the laboratory, as well as observation in the field, more than we have in the past to examine tenets we arrive at deductively.

The deductive theory of erosion must explain valley formation in terms of the laws governing the motion of running water; but, let it be added at once, only inasmuch as valleys are the result of incision by running water. The shape of valley-side slopes, and therefore the width of valleys, is the work of weathering and denudation, not incision, and must be studied in its own right. Initially, we need assume only that erosion seldom works on its own; it usually sets in motion the processes of weathering and mass-movement. We must therefore expect to find that the landforms to which erosion should give rise are seldom those we actually observe; they usually change straightaway to forms arising from the combined work of erosion, weathering and denudation.

The first question we must ask is, what is the nature of the erosive process? Even this has not yet been answered with certainty, and there is a lot more that remains still doubtful. Generally speaking, the theory of erosion is based upon the active force of uniformly-flowing water, though Brunhes attributes a major role to eddying, his *erosion tourbillonnaire*, not only in exceptional cases, where it is generally recognised, but in all cases.* The erosive power of a river changes from time to time with the volume of water discharged. High water seems to be a critical factor for the overall result; but the discharge during the rest of the year must have some influence. Studies of this would be of geographical importance too, since the annual flow régime varies from one climate to another and from one rock type to another. A difficult question, theoretically important, is whether the water works at and reshapes its bed by its own movement or only, as has been asserted, with material carried along on the bottom. Powerful erosion at Niagara Falls, only a little way downstream from the St Lawrence where it leaves Lake Erie, or at the Rhine falls at Schaffhausen, only a short distance downstream from the Lake of Constance (in other words, shortly after the river has deposited its material in the lake) argues against the exclusive effect of material carried along.

DIRECT, INDIRECT, VERTICAL AND LATERAL EROSION

Naturally, a layman thinks of a river as striking against a mountain chain like a ram's head in order to break through it; but this view sometimes echoes through scientific works as well. That it incises from the surface downwards is nearer the truth. It is often taught that erosion begins at the lower end of a river and progresses gradually upstream, where it is therefore only indirect. This theory was developed by Lyell for the deep gorge downstream from Niagara Falls, and is correct for tabular and plateau landscapes. Here rivers crawl sluggishly along a flat-floored valley, only to give way suddenly to erosion and plunge into a deep valley; sometimes old valley terraces can be recognised within these deep valleys preserving the former plateau course of the river. But in mountains, where the original surface is steeply inclined, direct erosion can set in simultaneously over the whole slope: a river must incise wherever the surface is inclined more steeply than its profile of equilibrium, in other words, wherever it can still carry along débris removed from its bottom or delivered to it from the side. According to the way it progresses, we can distinguish between direct and indirect erosion.[15]

Since flowing water gnaws at the bottom of a river bed as much as at its sides, the river can incise vertically downwards as much as sideways, and, while continually deepening its bed, move laterally at the same time. The view is often held that these two processes are consecutive, that the river first erodes vertically, and that lateral erosion begins only when vertical erosion has come to a standstill, the profile of equilibrium having been established. Lateral erosion will then be particularly noticeable because the level at which it works will remain constant, and a wide valley bottom (*Talsohle*)* is formed; but a river has to work laterally even when it is incising, and cannot avoid doing so from the start. Lateral erosion ceases only where a river has a perfectly straight course, since the river's line of maximum velocity then lies in its centre; where there is a curve and the thread of the current moves towards one bank, the river bites into the bank and enlarges

The Origin of Valleys

the curve. This lateral erosion combines with incision to cause the river to cut obliquely and not vertically downwards, as the contrast between the outer, steep and undercut bank of river curves and the flat slip-off slope of their inner side clearly reveals.

Erosion consists of the river tearing its bed and transporting the broken fragments together with the débris provided by weathering and denudation. Certainly during the period of low water, and on the inner side of curves in its course where it is sluggish, a river deposits material at the same time; sometimes the material deposited even exceeds the material transported. But where its power to transport material is sufficient to carry along more débris than it contains, a river erodes. Therefore the power of a river to erode depends principally on its power to transport, and the theory of erosion is on that account founded on this aspect of study.

But the theory cannot be content with the general distinction between a greater or a lesser power to transport material. It must go on to distinguish between the size of the individual fragments transported and the total mass of transported material. The size of the fragments depends mainly upon the gradient, the total mass on the active power of the whole river resulting from the combined effect of gradient and total volume of water. Large boulders will be moved only in rapid mountain waters, whereas only sand and mud will move on the bottom of gently-flowing large streams; but the total mass moved by such rivers is greater than the total amount moved along by mountain rivers as rocky material. A small mountain river must have a steep gradient if it is to transport its débris; it begins to erode as soon as the gradient is no longer sufficent for this to happen. By virtue of its volume of water, a large river can, on the other hand, continue to move sand and mud. It does not matter if single large blocks fall from the sides of the valley and remain lying where they have fallen; in the course of time they will be ground down and broken up.

Where there are local obstructions such as particularly resistant rocky banks or the talus cones of torrential streams (*Wildwässer*), incision is temporarily halted. Gradually obstruction will be overcome, and the river again incise until its fall is just enough at a given discharge to carry away the débris put into it. Thus it does not cease to attack and incise into its bed. But erosion in this case just balances deposition, with erosion on the outer side of river bends offsetting deposition on the inner, and the two processes alternate in the course of the year, so that at the period of low water the material eroded at high water is again deposited. So frequently are we confronted in the field by the fact that a river's gradient varies inversely with its volume of water, that we look for a special explanation of every exception. From this it follows not only that the theoretical deduction is correct, but that a river's course generally deepens itself rather rapidly to a state approaching equilibrium. This happens at different tempos and at different places along the course, according to rock resistance and the supply of débris, with the result that the state of equilibrium is reached sooner at one place than at another. But in the end equilibrium will be attained throughout the course of a river although, as every river engineer and boatman knows, the velocity of even sluggish rivers varies continually, probably as a result of irregularities in their beds.

THE PROFILE OF EQUILIBRIUM

The theory of the profile of equilibrium has long been known to hydraulic engineers and, as Baulig recently noted, was especially discussed by Surell in his study of alpine mountain torrents. Later Powell and Gilbert developed it using the expression 'base-level of erosion', a phrase also used by Heim. It was given a particularly precise form by Philippson, who introduced for it the name 'end-state of erosion' (*Erosionsterminante*).*

The position and shape of the profile of equilibrium is of very great importance to the understanding not only of valley formation but also of surface degradation as a whole. We must therefore look at it rather more closely.

Opinions differ on several points. Some researchers think that a river's longitudinal profile is suspended between its source and mouth, so that in the course of erosion the form of the profile changes, but not its height-range. This is theoretically impossible and we cannot use it to explain such surface features as cols in a mountain crest. The mouth of a river and thus the lower end of its profile, or to use Heim's terminology, its base-level of erosion, remains fixed unless uplifted or depressed by tectonic processes, whether at sea level, an inland lake, or a place where the river dries up or seeps away. Its source on the other hand, and thus the upper end of the river profile, is not fixed; it can be lowered by erosion in the course of time as the profile becomes flatter. In the state of equilibrium the source is at an altitude determined by the shape of the profile of equilibrium and the position of the base-level of erosion; it is independent of the original surface. When it is lower than this surface, the river can incise further headwards and incorporate more and more land into its course; how important this headward erosion is will be seen when we consider drainage patterns (*Talnetze*). If, on the contrary, the upper end of the profile of equilibrium is higher than the actual surface, as can be the case in tablelands, the land is spared from river erosion.

Since the volume of a river generally increases in a downstream direction, thanks to the excess of rainfall and seepage over evaporation and to tributaries, the gradient can decrease in this direction. The profile of equilibrium is therefore generally concave. But it is different when the river enters a dry area where it no longer receives tributaries, and evaporation exceeds precipitation. Its discharge of water then decreases, and a steeper gradient is needed to give it the same transporting power. The profile of equilibrium will be convex, and when the river dries up entirely, begins not at sea level but at some height either above or below it.

A further question is whether the profile of equilibrium is in fact a terminal state, whether in reality erosion ceases completely, as Philippson earlier supposed. I decided against this, in my study of Saxon Switzerland, as did A. Penck in a lecture on the ultimate limit of erosion.* On the one hand, the drainage basin can be enlarged and the volume of water thereby increased; on the other, the progressive degradation of slopes will reduce the supply of débris. The river, forced to handle a smaller load, has an excess of energy to lift and move material from its bed, in other words to erode and flatten its longitudinal profile and cause it to approach the horizontal asymptotically. That is in fact Davis's view as well. But it is no more than a deduction to use this to explain landscape planation; as far as I am aware there is still no inductive study of this, a study that must be linked with the study of valley bottoms. It is also questionable whether this near-horizontal flattening is more than a theoretical possibility, and whether the ground remains at rest long enough for such a state to be reached.

In a state of equilibrium, a river works only laterally and does little in a vertical direction. The valley bottom coincides at first with the high water river-bed, and in rivers with large periodic changes in water volume this may be an impressive terrace into which the low water valley bottom is incised. It will be widened by lateral erosion first in isolated places, especially in weak rocks, but gradually along its entire course. River windings or meanders, as they are usually called after a river in Asia Minor, seem to play an important role. We cannot, as did Honsell,* attribute them to a river's effort to slacken its course. Such an effort, a teleological concept, would manifest itself more readily in rapid rivers, whereas, in fact, only slow-moving rivers (those more or less in a state of equilibrium) take on meanders; rapid rivers, on the contrary, lack them. The question of whether the initial development of meanders depends upon the river being deflected by some small obstruction, or starts with some kind of rhythmical fluctuation of moving water about the stream's axis, can be left open. But once there is a deviation from a straight course it will be progressively enlarged by the more intense erosion on the outer side of curves; once a bend develops it seems to result in the development of further bends and the entire river acquires a meandering course. The amplitude of the meanders corresponds with the size of the river; the meanders are not stationary, but migrate slowly downstream. In this way, the ground between them is levelled and a continuous, broad valley bottom gradually develops, its width corresponding to the amplitude of the curves. When this is less than the width of the valley floor (*Talböden*) it probably indicates that the river has atrophied as a result of climatic change or capture. We must not reject the possibility that the river, in 'tottering' (to use Spethmann's terminology) back and forth across the valley as it sometimes does, extends the valley bottom beyond the amplitude of the meanders, or that the river's axis may shift to one side. But the causes given for this—the influence of the earth's rotation, side-slipping across weak beds, and other explanations—are not really plausible.

A substantial widening seems generally to occur only as a result of deposition, either when the land sinks relative to sea level or to the foreland, or sea level rises and the river is ponded back, or when its volume of water decreases, or its load increases.

Although as a consequence, broadening of the valley bottom proper is limited, the foot of valley sides can evidently be flattened so much that it appears to virtually blend into the valley bottom. Only a small change in level or even an exceptional high water may be sufficient to

The Origin of Valleys

incorporate the two. Perhaps we could use the expression valley floor in this wider sense as distinct from valley bottom. But the width of such a tray-shaped valley floor will in fact always remain limited. We cannot assume without evidence that broad flats have been valley floors.

RIVER AND VALLEY MEANDERS

River meanders are to be distinguished from valley meanders where the valley itself winds and not merely the river in its valley. This results in interlocking spurs and, when breakthrough later takes place, the cut-off meander cores. That valley meanders have fundamentally the same origin as river meanders can hardly be questioned. The question is whether they develop as the river incises when, for example, it crosses harder rocks, as Georg Wagner assumes is the case in the Nagold valley,* or are established on a flat valley floor and are only enlarged during incision. Very early on Ramsay* advocated this latter view. Many others, myself included, have accepted it, and it has become more or less of a dogma for Davis's school. Davis sees valley meanders as evidence that rivers formerly flowed in a profile of equilibrium and are now in a 'second cycle'. Behrmann even thinks that the incision leads to a reversed development, the now rapidly-flowing river striving to re-straighten its course.* Only observation will resolve the question and this has as yet hardly begun. If the meanders first developed during incision, the upper edges of the slip-off slopes will together form a more or less straight line. If on the other hand they were already established before incision, the upper edges must meander. There has as yet been little study of either the relationship between valley meanders and the direction of the strike of strata, that we can see so clearly in the Rhine Massif, or of that between valley meanders and rock type, to which Bach referred as early as 1853, in the case of the Muschelkalk, and which Wagner tried to explain as the result of greater resistance to vertical incision.*

In theoretical discussions of erosion, internal build and the tectonic surface are usually taken to be constants. This is not of course a return to the Catastrophist's theory, with internal build being completed in one mighty upheaval, as though in a single day; its development is recognised as a slow process. But nevertheless it is thought of as a process taking place at a speed greater than that usually ascribed to erosion and denudation. Internal build is seen as occurring first, while erosion and denudation follow. Only in certain exceptional circumstances, to which I shall later refer, is it regarded as keeping equal pace with them. Recently W. Penck has opposed this view. In his opinion erosion and denudation usually and without exception keep pace with mountain uplift, with the result that the tempo of uplift is expressed in the degree of degradation, especially in the bevelling of slopes. He maintains that gentle slopes correspond with slow uplift, and steeper slopes with more rapid uplift. If a valley side is steep at the top and flat at the bottom, *i.e.* concave, uplift has slowed down. When it is less steep at the top and steeper at the bottom, *i.e.* convex, uplift has speeded up. In the former case, he speaks of waning development, in the latter of waxing development; for him the form of slopes is diagnostic in assessing the intensity of uplift. This is the inspired and shrewd speculation of a young research worker who died too soon. But is it more than a speculation? Will deduction be supported by inductive evidence? Are not the various slope forms much more closely related to the kind of denudation resulting from rock type and climate? At all events the hypothesis is at present not strong enough to carry the edifice Walther Penck built upon it.

CHAPTER III
Valley Terraces

Valley terraces are among the most important characteristics of the shape of many valleys. Forming horizontal or gently-inclined breaks in valley-side slopes, they are sometimes no more than a ledge; in other cases they are of considerable breadth. In some instances they are a prominent feature of a valley's physiognomy and apparent at first sight, in others recognisable only to the practised eye. But valley terraces are always of great significance in the origin of valleys. Terraces which extend along the sides of valleys and determine their cross-profile have been termed lateral or side-terraces to distinguish them from the stepped longitudinal profile of many valleys, especially those of the Alps. Here we are concerned with the former alone; even they are of very different kinds. Here we must use the diagnostic method, in Passarge's medical term, 'differential diagnosis',* as a method of study.

A twofold difference can be observed among valley terraces. The first involves their composition. Many are formed in the bed-rock of the valley side and carry little or no gravel upon them; these are rock terraces. Others, in contrast, are built of great thicknesses of gravel descending to the valley floor or at least resting on bed-rock only at some considerable depth; these are gravel terraces. A second difference observed among rock terraces is more difficult to interpret, but genetically still more significant, and of first importance in distinguishing rock terraces from gravel terraces. Some rock terraces, often covered by gravels, run roughly parallel with the valley floor, rather like the surface of gravel terraces. But unlike gravel terraces they truncate the strata, unless by chance the strata have the same low inclination as the valley floor, and show no relation to them. The terraces of Alpine valleys, as Rütimeyer has shown,* and the terraces of the Rhine valley in the Rhine Massif, are of this type. The other kind has nothing to do with the valley floor, but is more or less markedly inclined in the opposite direction and generally coincides with particular beds or other structural surfaces. They are extraordinarily common on a small scale, less frequent on a larger one, since they can occur on a large scale only on more moderately inclined strata beds. An example is the terrace of the Neckar Valley at Heidelberg which sinks in an upstream direction. Unfortunately, we have no terminology for the two kinds of rock terraces based upon descriptive characteristics alone; but they are manifestly of different origin, and we can therefore employ genetic terms straight away. The first kind are old valley floors, corresponding to a halt in erosion and the formation of a wide valley floor; so they can be termed terraces of erosional stillstand or simply erosion terraces. The other type have nothing to do with erosion, but are developed by weathering and denudation and therefore called weathering or denudation terraces. In most cases the two kinds can be readily distinguished; but they can sometimes be mistaken for one another. When the strata lie roughly horizontal, denudation terraces will accordingly be nearly horizontal and not unlike old valley floors. This led Dutton to see the Esplanade at the upper edge of the Colorado canyon as an erosion terrace or an old valley floor, while Davis has recently explained it as a denudation terrace. On the other hand, steeply-inclined terraces can be interpreted as old valley floors if subsequent tectonic disturbance is accepted. Thus Heim considered the terraces of Lake Zürich, which fall gently in an upstream direction, as old valley floors later uplifted. His view that they were the result of foreland uplift was based upon this. Brückner thinks that they are associated with particular strata, in other words are denundation terraces having nothing to do with uplift.*

So we must distinguish three classes of terrace: weathering or denudation terraces, old valley floors or erosion terraces, and gravel terraces, the surfaces of which are also old valley floors but of a different kind.

THE DIFFERENT CLASSES OF VALLEY TERRACES

Denudation terraces owe their origin to variations in the rate at which weathering and denudation degrade the side of a valley. This variation is the result of differences in the resistance of horizontally-bedded rocks to these processes. This is less a question of hardness in the physical sense of the word than of their permeability to water and the type of denudation. When beds dip steeply, or when different rocks stand vertically alongside as with eruptive dykes, terracing does not develop; instead there is an alternation in the longitudinal profile between protruding steep rock bars and receding gentle slopes, as can be clearly observed, for example, in the valleys of the Rhine Massif. True terracing develops only when the beds are more or less horizontally arranged, with their most frequent cause the outcrop of a spring line above impervious rock.

The nature of terracing varies. Many terraces become very broad, while others are no more than small ledges in the slope. Sometimes many terraces lie above one another. The rock faces of Saxon Switzerland are often most strikingly terraced in this way, probably because the frequent small variations in rock composition favour groundwater emergence and the undermining of rock faces. In other cases, when, for example, thick masses of limestone or sandstone overlie clay beds, only single large terraces develop. The terracing of the Swabian Keuper, with its repeated alternation of clay and sandstone, occupies an intermediate position. Terraces are especially well defined on a valley side where horizontally-bedded strata overlie remnant surfaces in the basement complex. The former are usually more intensively degraded than the latter and therefore form the parts of the valley sides which have receded furthest.

A special kind of denudation terrace, one to which Richter has drawn attention,* is found at the upper edge of the sides of many Alpine valleys. They appear to have been formed by the recession of corries. They are not associated with any particular rocks, but are related to an altitudinal climate zone. Many research workers see them as old valley floors.

If it is true that the mechanism by which denudation terraces are formed is not yet entirely clear in its details, there is a much greater need for detailed individual studies of erosion terraces. Erosion terraces are rock surfaces usually overlain by gravel. Geographers give most of their attention to the former, geologists to the latter. Geographers are more at home and practised in analysing landforms, geologists in working with a hammer and examining the rock. But too strict a division of labour is unhealthy and leads to incomplete results. It seems to me sadly one-sided for a research worker to consciously restrict himself to one of the two methods of investigation; assured knowledge is to be gained only by combining them. A geographer working as a geomorphologist must acquire the ability to distinguish between different kinds of gravel, and, within limits, to determine their source; in difficult cases he may call on the help of the geologist. Conversely, the geologist, who in his survey work also takes account of terraces, must acquire a sense of landforms and familiarise himself with the law of their development; otherwise it is only too easy for him to mistake any gravel lying on a slope for a separate terrace, or rock terraces for gravel terraces, or fail to distinguish between terraces developed at only one site, for whatever reason, and the extensive continuous terraces, which are in fact old valley floors.

Such valley terraces were first studied by Rütimeyer in the Reuss and Ticino valleys, and recognised as the remains of old valley floors formed by the river during a pause in mountain uplift. When downcutting was renewed, the river incised itself into its floor, headward erosion shifting rapids (*erosionsstrecken*) or knickpoints (*Gefällssteilen*, after Sölch) upstream. It can be left open whether his interpretation is entirely correct; but he pointed the way to further study and understanding. Its importance in the developmental history of mountains was recognised, and we began to speak of periods of erosion. When Davis introduced the phrase 'cycles of erosion' he was not introducing any new factual knowledge, merely another and hardly better expression to suggest a series of recurrent changes.

Valley bottoms, formed by the displacement of a river's course when its profile is in a state of equilibrium, must be horizontal in cross-section and sharply demarcated from the valley-side slopes. But in valley terraces, we are confronted not only by actual valley bottoms but by valley floors in the wider sense of the word, which include the quite flattened foot of the valley side. And they have subsequently been further smoothed by the accumulation of débris at their inner edge and its removal from the outer edge, although here a perceptible break of slope always remains. Because of this smoothing process it is generally difficult to determine the original height of the terraces. This is often taken without further examination to be the height of the lower edge of the present smoothed surface; but in fact it usually needs to be sought at a higher level than this. In most Alpine valleys there has been additional remodelling and smoothing by glaciers; we can scarcely continue to regard their valley terraces as features of fluvial erosion.

Valley Terraces

Old valley floors are never preserved in their entirety. They are not only interrupted by tributary valleys but also often fade out on the outer side of valley curves where they have been destroyed by erosion advancing into this side. We must therefore reconstruct old valley floors from a number of fragments, no easy task when we bear in mind the subsequent smoothing to which I have referred. This explains why there have been such varied reconstructions, for example that of the Reuss by Rütimeyer, Heim, Bodmer, Brückner and others.* Every piece of erosion terrace must once have been valley floor; but it need not always have belonged to a continuous one traceable over a long distance. Indeed in present-day valleys, even when there is no general valley floor, we often see a flat area where the valley is wider at particular places, such as on the inner side of river curves, or upstream from constrictions. Such flat areas may survive as local terracings when the river incises. In the Moselle valley below Trier, there are only a few terraces which are continuous downstream; but between these continuous terraces there are numerous local terracings. And in many other places the number of terraces is much smaller when we make this distinction. Only the large continuous terraces which extend beyond the main valley up into tributary valleys correspond to periods of general erosional stillstand and are of greater interest in the developmental history of mountains.

The surprisingly great width of the old floors of many Alpine valleys, as it is reflected in the distance between terraces lying on both sides of a valley, was stressed by Richter in his geomorphological studies in the High Alps.* But this is in part illusory; the old valley floors were winding, and it was precisely the outermost curves that were more likely to be spared from destruction during incision. But if these outermost fragments are linked directly with one another they give the impression of an inordinately broad valley. Only detailed studies, such as are perhaps only rarely possible at present, can determine an old valley floor in detail and give us an accurate judgement of its width. Sölch's belief that the older valley floors are, the further they can be followed upstream, requires more examination.* Many levels which have been taken by some workers to be old valley floors have been thought by others to be more like corrie platforms.

In the case of continuous terraces it is always possible to recognise the alternation of a period when the valley floor was widened by lateral erosion and vertical erosion ceased, and one when vertical incision was renewed. In any event such terraces always indicate a change in a river's power to erode. This could result from an increase in water discharge or a reduction in load—in other words, from a change in climate. But in a moist climate with only a moderate fluctuation in rainfall, climate changes are scarcely great enough to lead to terrace formation. And it is exceptional for a river's water volume to be increased by the capture of a river hitherto independent. Rütimeyer's explanation, the foundation of Davis's viewpoint, has been pretty generally accepted and is on the whole correct. He considers that a new phase of erosion is initiated by re-uplift, and consequently that erosion terraces are to be seen as indicators of intermittent uplift. Erosion terraces traced over wide areas are the best if not the only means of establishing the form of general uplift or upwarping, and whether it is limited in area to individual mountains or extends to areas of different structure, when this can no longer be seen from the way strata are disposed. The study of erosion terraces has for that reason become an important working method in geotectonics, but surprisingly enough only in recent years.

GRAVEL TERRACES

The level rock surface of gravel terraces is covered not by just a thin layer of pebbles, but by a great thickness of gravels. Their level surface indicates that they too are old valley floors cut by rivers. But in this case, the valley floor can not have originated solely by lateral erosion during a halt in incision; it must have been formed by deposition, the nature and cause of which is, however, debatable. Penck* was the first to study the extensive and very thick glacial gravels of the Alpine Foreland and Alpine valleys, especially that of the Inn. He regarded these gravel accumulations as a result of the increased provenance of débris by advancing, strongly eroding glaciers, and the inability of glacial streams to carry away their load. In the phases when less débris was provided, that is in the interglacials and the postglacial periods, the rivers again incised into the gravels. Although he later changed his explanation for the Mittelgebirge of the Inn valley, it will certainly be found to be right in many instances. When Sievers* found extensive gravel terraces in the Cordillera of Mérida he unhesitatingly applied Penck's explanation to them, since he also found evidence of former glaciation. But he did not observe the caution always needed when an hypothesis is transferred from one locality to another. He neglected the fact that the terraces extend downstream much farther than glacial débris can have been transported, and that here they even attain their greatest thickness. I observed the same kind of gravel terraces even more massively developed in the Cordillera of Bogotá, near Fusagasugá, between Tocaima and La Mesa ('the table', so named from its position on such a gravel terrace), and at many other places.[16] Here they usually occur where any link with old glaciers is impossible. Only exceptionally are they found in high regions; in the overwhelming majority of cases they occur at medium and lower heights from around 1,900 metres well down into the Tierra Caliente. In the valley of the Río Bogotá, especially, between La Mesa and Tocaima, several systems of gravel terraces can be distinguished. The lower ones do not appear to have been cut out of the upper, at least not always; they seem to be the result of renewed deposition, indicating that two periods of deposition have interrupted the development of the valley. Changes of climate, in the form for example of intervening dry periods, could be invoked to explain this; but since the terraces occur only in transverse valleys and generally lie upstream from defiles (*gebirgsriegeln*), they are more probably the result of river damming by uplift or folding. Unfortunately I did not have the opportunity to investigate whether the gravel terraces continue beyond the edge of the mountains; this must be the case if they are to be explained by changes of climate. Identical, or seemingly identical phenomena can easily have different causes. An explanation which has been proven in one area should not be transferred uncritically to another; instead, renewed painstaking individual study is always necessary.

In keeping with their different origin the three types of terrace have different distributions. Denudation terraces occur where rocks of differing resistance lie on top of one another. Erosion terraces appear to be associated mainly with renewed mountain uplift after periods of erosional stillstand. Gravel terraces depend upon changes in the energy of flowing water and result from changes in gradient as much as from changes in water volume. They are therefore just as easily the outcome of tectonic as of climatic changes, and can therefore be found in very diverse localities.

A thorough study of valley terraces opens up wide perspectives for the understanding of landform development as a whole. Denudation terraces are an early phase of the great benchlands (*land-terrassen*)* we see in scarped landscapes.¹⁷ Erosion terraces can be the first stage of remnant surfaces or peneplains (*fastebenen*) in the Davisian sense. And gravel terraces occur repeatedly on a grand scale in dissected upland surfaces, such as the Swiss or Swabian–Bavarian plateaux; they help one to understand what Klupfels meant by a 'boxed' landscape (*Schachtellandschaft*),* where the valley forms and surface features of different periods lie within and above one another. It is often difficult to distinguish the several kinds of extensive regional levelling, as we shall see. But a principal criterion for doing so is whether they are coupled with denudation or with erosion terraces.

CHAPTER IV

Age and Form of Valleys

AGE AND STAGE OF DEVELOPMENT

The idea that different parts of the earth's surface are of varying ages, that they have been subjected to the effect of modifying forces for different periods of time, and therefore display dissimilar characteristics, is not new. Research could not but progress beyond the stage at which the theory of development superseded the catastrophic viewpoint. Since the actual surface of the earth is not the surface which internal surface build would of itself produce, not the tectonic surface, but that derived from the tectonic surface through the impress of external or exogenic forces, and since such processes are not catastrophic events, but the immense effect of long-continued modification and destruction and the summative effect of lesser forces still operating on the earth's surface, the degree to which the surface is modified and destroyed must depend upon the length of time which has elapsed.

Given this concept, it was not long before research workers tried various ways of dating the modifying forces, and thereby the landforms they shaped. English researchers, especially Lyell and de la Béche, observed the rate at which coastal cliffs retreated. Lyell tried to measure the retreat of Niagara Falls, and in this way determine how long was needed for it to carve the gorge below the falls, and cut back to Lake Erie. In Switzerland, Rütimeyer and Heim studied the duration of sedimentation processes, for example the formation of a delta by the River Reuss in Lake Lucerne. The Norwegian, Kjerulf, wrote a monograph on the 'chronometer' of geology.*

Recently, the Swede, de Geer, determined the length of time since the glacial recession by counting the layers of so-called varves, as well as the constriction of *osers*.* And datings have been attempted by relating geological processes and events such as the Glacial Epoch to particular astronomical conditions. But except for the studies limited to the recent geological past of de Geer and his successors, attempts at dating have not as yet been very fruitful. The age of prehistoric processes, or at least of those earlier than the last recession of the northern ice, can still be determined only relatively to that of other processes and states, especially to the occurrence of characteristic animals and plants. We can still do no more than arrange them within geological periods or formations, giving them a geological age.

The concept of age has another implication too. Since the modifying processes are persistent and reshape the earth's surface progressively, the landforms they create must continually change, and differ more and more from the original or tectonic surface, until in the end this surface is completely destroyed. We can therefore distinguish stages of development and set up a series of developmental stages. But such modification and destruction does not proceed at the same tempo everywhere; it is faster or slower according to the energy of the processes and the amount of resistance they meet. In keeping with the time needed to attain a given stage of development, its absolute and geological age will differ from that of the same stage elsewhere. In many instances we can ignore differences of duration as being almost irrelevant and use only the stage of development.

This way of thinking is not new either. Throughout the literature we can find comments pointing in this direction, since Desmarest, in his studies of the volcanoes of Auvergne, interpreted their various forms as the result of their being at different stages of modification.*

In the study of valleys the concept took on a particular form as the doctrine of the profile of equilibrium, in that steep valley bottoms of irregular gradient appear to be in a primary stage of development, and flat and uniformly sloping ones in a more advanced stage. As escarpments are formed, we can follow the development of individual valleys to the stage at which, by incising headwards, they destroy the continuous surface until only a number of residual masses of it (*Zeugenberge*) remain. Moreover, the concept was important for our understanding of whole mountain systems, indeed of landscape in general, once the major role of degradation had been recognised. Heim, and later others too, showed that contrary to the older view that high mountains were the oldest features, only youthful mountains can rise to great heights, while older ones are of necessity degraded. The stages by which different types of mountains are modified have been described. Gilbert did this using the Henry

DAVIS'S INTERPRETATION OF AGE

But our knowledge of surface features was based primarily on the study of internal build and the nature of modifying forces; study of their age or stage of development took second place. Davis, on the other hand, has since 1884[18] put stage to the forefront. 'Time', he says in his 1899 essay on the geographical cycle,[19] 'thus completes the trio of geographical controls—alongside internal build (structure) and the kind of surface modification (process)—and is, of the three, the one of most frequent application and of most practical value in geographical description.' And in his *Grundzüge der Physiogeographie*, stage is to the fore. Any research worker from his school defines the character of a landscape primarily in terms of stage and lays most emphasis on this in geographical description. We must dispute this approach because it has gained great influence.[20]

Davis compares the development of a sequence of landforms, or as he puts it 'a cycle', with a lifetime, and compares the stages of landform development with those of human life. When the land surface is sculptured in a moist climate principally by running water, moist weathering and rainfall, that is to say in his 'normal cycle', Davis compares the land's emergence from beneath the sea and its uplift to a specific height above sea level to the so-called primary stage of childhood, while subsequent stages of development are termed youth, early maturity, maturity, late maturity, and old age. The last is typified by the wasting of all protuberant landforms and their almost complete levelling to a stage resembling that of childhood but at a lower altitude. The basis on which stages of development are compared with one another is their degree of development. A landscape in the stage of childhood is like a child's character, uniform and undifferentiated. The surface of the land, like a living being, gradually becomes more and more complex. Just as when mature, a human life displays the greatest development of its powers and individual traits, thus the land goes through a period in which landforms are most amply (*reichsten*) developed, dissection furthest advanced, and rivers have finished their vertical incision. Thereafter, living things begin to decline in power, to degenerate and, within limits, even to become smaller. In a similar way the earth's surface forces become less intense and landforms become progressively less prominent until finally obliterated completely; the landscape is again childlike. But renewed uplift can initiate a new cycle and reinvigorate the forces.

Allow me first a linguistic and methodological comment!

A comparison of processes of inorganic nature with those of life can serve as a temporary sketch of an idea, animate presentation, and be beautiful and effective. But when it recurs again and again, it becomes stale and loses its point; when used excessively, it is merely tasteless. I can imagine, as did the authors of folk tales, that an old person may suddenly become young; but the idea that a young or mature person is etched out of an old or senile one is out of the question.

It is characteristic of primitive peoples that they compare things with life in a mythological fashion. Science has gone beyond the idea of an animate or spiritualised inorganic nature; it seeks to know the processes of inorganic nature as such, and where possible to trace the processes of organic nature back to them. Thus science progresses in precisely the opposite direction. If it uses metaphor and bases its terminology upon it, it runs the risk of seeing metaphor as explanation. Content with this, it may fail to analyse individual processes into their physical and chemical components or to study rigorously their causal interrelationship. I cannot help feeling that it is to this that Davis and his followers have succumbed. Without showing how they are linked together causally, this school of thought assumes that the different traits or form characteristics of valleys, or indeed of the land surface as a whole, change simultaneously and harmoniously.

Davis's designation of surface features by age is, moreover, ambiguous and equivocal. His terms are not only supposed to be of chronological value in that they date the features, but to express degrees of development as well. Yet development takes place during a specific period of time; it is necessary to say how long the development took and to consider whether the same stage of development is reached in the same time everywhere or takes a longer or shorter time in different cases. But on the one hand Davis talks about time, on the other about stage.[21] We have to draw two conclusions from this inconsistency: firstly, the knowledge that the stage reached (*Ablauf*) is not proportional to the time that has elapsed; secondly, the great importance attached to time for the progress of landform development.

In reviewing Dietrich's work,[22] A. Penck saw as especially important the evidence that in the Moselle valley the different age-stages in different parts of its course had apparently formed at the same time, in other words independently of the time which had elapsed. For this reason he would distinguish between morphological age and the actual or geological age. In so doing he is thinking of the differing rate at which men, and even animals, age; as he once put it, at fifteen a girl is young, but a dog is old. Davis himself adopted this idea: an old mushroom beside a young oak tree is nevertheless younger than the oak. But we cannot assign different ages to the several parts of one and the same living thing! A man cannot have young arms and old legs; nor can one and the same section of a valley have both young and old landforms. If we use descriptions of its age to characterise a landscape, we are expressing the time it has taken to develop; we must distinguish its stage of development from this, since the speed with which it develops depends on the intensity of the destructive forces and on rock resistance as well as on the time that has elapsed.

Age, taken to mean stage of development, is, therefore, a complex, not a simple term; it combines several unknown components. To designate landforms by age used in this sense is no explanation at all, but a purely empirical description, and this is precisely what Davis himself attacks; but curiously he and his pupils have not realised this. If it is to become an explanation, stage of development must be broken down into its two components: the time involved and the tempo of development, which is in turn controlled by two factors, the effective force and the degree of resistance.

The speed with which different stages of development succeed one another calls for impartial investigation. It is not so much the number of years needed that we want to know. It will be a long time before we can establish this other than in exceptional cases. What we want to know is how these different stages are related to geological periods. Opinion still differs widely as to how long each stage requires. Whereas it was thought at first that a valley needed a long period of geological time to develop to the stage at which its longitudinal profile was in equilibrium, and that an even longer period was needed for the complete planation of a landscape, these are now often compressed into short time-spans. It is becoming more and more plausible that many valleys have been incised only since the glacial maximum (*Haupteiszeit*), as I established as early as 1887. But the later stages of surface sculpture—attainment of the profile of equilibrium, slope lowering to the extent of complete levelling, and most certainly the development of a whole drainage pattern—takes much longer than incision, as Davis himself declared. For all these processes to have run their course, not just once but repeatedly during short phases of Tertiary time, as ardent supporters of Davis maintain, they must progress at a much faster rate than is usually thought to be the case, or alternatively the geological periods must have been much longer than on other grounds we have thought it necessary to postulate.

According to W. Penck, degradation more or less keeps pace with uplift. For the concept of age he substitutes that of the faster or slower rate of uplift; rapid uplift produces youthful landforms, slower uplift old, indeed senile, landforms. Great as these differences of approach appear to be they are in fact closely related; both allow for only quantitative and not qualitative differences of process, and both seek to compress the overall form of a valley into a single term.

AGE CHARACTERISTICS

But more important than the question of how long developmental processes have been effective is whether the form characteristics ascribed to a given stage of development do in fact succeed one another, whether the later derived from the earlier, rather than originated side by side under different conditions of internal build and climate. Do they represent stages of development at all; or are they different kinds of development? Like Davis I will begin to try to answer this question with reference to valleys.

The first characteristic of a valley's age is its gradient or longitudinal profile. In its initial state, and in the developmental stages of infancy and youth as well, this will be controlled by the tectonic surface and will therefore vary with internal build. If the tectonic surface has an appreciable overall slope, as is probably the case with most volcanoes and fold mountains, rivers can incise immediately throughout their course (*see* p. 20). If, on the other hand, the tectonic surface is flat and forms a scarp only at its margin, as in tablelands and remnant uplands (*Rumpfgebirgen*), rivers can at first incise only at this margin, and erosion shifts gradually headwards. Each stream course is comprised of an upper inert, tranquil stretch, and a lower one with a steeper gradient and faster flow; the bigger the stream the

Age and Form of Valleys

faster the lower course extends upstream. In both cases, whether erosion begins simultaneously along the whole length of the course or progresses gradually headwards, the gradient and stream velocity depend partly on the difference in height between the crest and the foot of mountains. But we must not let our experience in Alpine conditions deceive us; the incompleteness and ruggedness of Alpine valleys is not only a result of the mountains' altitude but of their former glaciation as well; it is not found in this form in unglaciated high mountains. The swiftness with which incision takes place depends also on the volume of water and therefore, indirectly, on climate and soil. It depends on rock resistance too; where hard rocks adjoin weak ones, stretches of steeper gradient alternate with ones of lesser gradient. Thus in their youth rivers have a variety of forms; their only common feature is their incomplete nature and their dependence on structure and the tectonic surface. But they are working towards a common goal, namely the profile of equilibrium, and this is controlled only by the volume of water discharged not by the tectonic surface.

The profile is flatter in large rivers, steeper in smaller and therefore particularly so in all headwaters. The tempo with which the profile of equilibrium is attained varies from case to case, but is a particular stage of development. After Davis we can term this the state of maturity, for linguistically the word 'mature' carries no implication of time; rather does it connote a state, and seems less questionable a term to me than one conveying age. But we must be on our guard against linguistic vandalism; we must use the word 'mature' carefully. We should speak of maturely-incised valleys, instead of valleys incised to maturity. As we have seen (*see* p. 23), the profile of equilibrium when attained is not theoretically a constant. Since the amount of material provided by degradation decreases, the water needs less power to carry it along, with the result that when the water-volume remains constant the gradient can progressively decline. But such old, or in Davis's terminology senile, stages of development are only theoretical; they have as yet been scarcely observed in nature. Moreover, it is difficult to use the longitudinal profile to distinguish them from the stage of maturity. The formation of a valley bottom is a second characteristic which helps us to determine the stage of development. So long as the river cuts vertically downwards it also cuts laterally; lateral erosion combines with vertical erosion to bring about lateral shifting and the elimination of entrenched meanders (*see* p. 24). Locally, especially where stretches of weaker rocks lie upstream from a rock bar, broad valley floors can be formed in even as early a stage as this. But, generally speaking, a wide valley bottom will be formed by lateral erosion only after vertical erosion has ceased. Thus, while isolated stretches of broad valley floor are no indication of the stage of development, a continuous valley floor which extends along the whole course of a river can be considered quite as good an indication that the state of equilibrium or maturity has been reached as a uniform gradient. It is likely that the width of a valley has nothing to do with its age, but depends on the size of the river (cp. p. 24).

The greater or lesser steepness of its sides serves as a third characteristic of the age of a valley. Young valleys should have more or less vertical walls—canyon form is explicitly attributed to recent uplift and the youthfulness of valley formation rather than to dry climate—whereas the flanks of mature valleys are moderately inclined, and those of old and senile valleys quite flat. Such an interpretation is based upon the valid distinction between the formation of vertical or even overhanging walls by river incision, at least along the straight sections of its course and on the undercut outer sides of bends, and the bevelling (*Abschrägung*) of slopes by weathering, gravity, rainfall and groundwater. But it mistakenly assumes that this distinction, correct as an abstraction, is a distinction in fact; it assumes that bevelling follows upon river erosion. Valleys are seen as being formed in two successive acts, namely one of incision and another of slope bevelling. In truth, bevelling begins at the same time as incision; whether or not it keeps pace with it, and the detailed form it shapes, depends on the depth of the valley, on the intensity of forces climatically controlled, and on rock resistance. Only secondarily, if at all, does valley formation depend upon age, that is to say on the duration of processes. In incoherent or weak material subject to surface runoff, steep or even vertical valley sides (sides such as Davis maintained should be present in the youthful stage) cannot survive for even the shortest period. Under the influence of gravity, weathering and degradation they are immediately worn down. And quite young valley sides display the supposed characteristics of maturity, and contrast sharply with the irregular, still-youthful profile of the valley bottom.[23] The young valley has the impress of 'age'. Even on hard but impermeable rocks, such as most crystalline ones, slopes are bevelled fairly quickly in a moist climate by surface run-off and soil creep. A canyon can never be produced under these conditions. Conversely, steep walls survive in a dry climate or in permeable rocks where there is no surface run-off, even when the valley bottom has a profile in equilibrium and has been widened by lateral erosion, as in most valleys of the Swabian Alb. In the course of time valleyside slopes may be flattened. But the minor role that age plays in this can be seen in 'polycyclic' valleys, where slopes sited above valley terraces and formed in an earlier period of erosion ought to be much flatter; in fact they are usually at the same angle in any one type of rock.

The density and ramification of the drainage pattern should form a fourth characteristic of valley age. In the youthful stage, drainage should be open, valleys only slightly ramified, and the mountain mass or highland correspondingly little dissected and intact over wide areas. As time passes, erosion and valley development should involve a greater area. The formation of valleys over an entire area, so that none of the original surface remains intact, is taken as a characteristic of maturity. We previously defined the state of maturity as that in which the larger valleys had attained the profile of equilibrium; this must therefore take place in the same time as is required for valleys to develop over the whole area. I can see no evidence of this being so, nor can I see any reason for such a close relationship between the two phenomena. The term 'mature' is ambiguous when founded on two different criteria, independent of one another. The acquisitive progress of erosion is by no means the same in all areas; it changes in regions of different structure. In volcanic landscapes and mountain chains, the whole land surface is immediately and rapidly attacked by erosion, and the appearance of maturity quickly taken on, while all rivers are still turbulent mountain waters, and long before the valley bottoms attain the profile of equilibrium. One need only look at a range of mountains which are not of limestone! On oldlands and tablelands, on the contrary, a long period must pass before erosion attacks the whole of their area; even after the larger rivers have long had a tranquil course, extensive plateaus or blocks remain undissected. The density and ramification of the drainage pattern depends on internal build and on the amount and régime of rainfall more than on the stage of development.

A fifth characteristic can be sought in the relationship of valleys to internal build. Insofar as they have not been incised by successful rivers during mountain uplift and have survived, or to use Powell's term, are 'antecedent', valleys will at first be controlled by mountain structure. In the course of time this dependence on structure is progressively reduced by the work of the rivers themselves; rivers favoured with a large volume of water or a considerable height range, or those on weak rock, encroach upon the area of less-favoured ones, and transgress watersheds sited in relation to tectonics. On oldlands and tablelands new valleys develop as well. Gradually, the valleys free themselves of their dependence on the tectonic surface and adjust to their conditions of work, on the one hand to their varying discharge, on the other to the hardness or overall resistance of the rocks. An abundance of secondarily-formed valleys can, in fact, be seen as a characteristic of advanced development. But it is difficult to associate this with particular developmental stages. We must always look carefully first to see whether the alignment of valleys is not actually controlled by tectonic conditions, for example by small faults and clefts, as is probably the case with many so-called subsequent valleys.

Thus the several characteristics cited differ in value as indicators of the age of valleys or their stage of development. The profile of equilibrium and a broad valley bottom, characteristics largely dependent on how long a period of time has passed, can serve as good indicators of a particular stage of development. The ramification of valleys and the degree to which they are related to the original tectonic surface are to a certain extent a result of these; but valley ramification and independence of the tectonic surface can scarcely be associated with the profile of equilibrium and with broad valley bottoms in a particular time relationship, and all of these features are used to characterise a particular age. Slope development, the principal control of valley physiognomy, only incidentally depends on the stage of development. Not only does flattening occur in many instances so early that Davis speaks of valleys being born mature, in others so late that one could speak of valleys being eternally young, but slope flattening is as a whole more a function of climate and of rock type and the way it is disposed than of the stage of development. It is a disastrous error to believe that the form of valleyside slopes can be characterised according to their age.[24]

THE PSYCHOLOGY OF ERROR IN ASCERTAINING AGE CHARACTERISTICS

If a scientific postulation is to be more than a fanciful idea it must be tested against reality; if experiment is impracticable, comparison is the means of carrying this out. Not for nothing did Peschel introduce the comparative method into geography. Our methodology will not progress beyond Peschel's by renouncing the comparative method but rather by using it more fundamentally, by using thematic maps as well as general ones, and still more by direct observation in the field. Passarge's demand that a geomorphological map should accompany every geomorphological study is too extreme, and not a practical proposition; nevertheless, it is fundamentally sound. Every geographical hypothesis must at least be tested in typical examples against the map. Before we settle conveniently into the conceptual edifice of an approach based on age, we must, by load testing, determine whether the girders will carry the building. We must draw maps for different areas of diverse build which incorporate the age of valleys according to the characteristics we have reviewed; longitudinal profile, width of valley bottom, the form of the valleyside slopes, the density of the drainage pattern and its relationship to internal build. Such a map would enable us to see whether these different characteristics correlated with one another in their distribution, or differed, whether and to what extent they are related to rock type and disposition and to climate; it would allow us to decide whether these features do in fact have common characteristics which transcend both rock-type and climate and can be attributed to age.

We should also examine whether the age-stages deduced theoretically actually do occur in nature, and whether there is any notable difference in the frequency with which each occurs. Do old valley forms, midway between maturity and senility, actually exist in any rocks other than clay or similarly weak, impermeable ones, when 'old age' obviously depends on rock composition alone? Do 'senile valleys', where one can recognise earlier stages of development, ever occur? Or is it not much more often a case of valleys being flat from the start or even no more than dells?

In contradicting Davis's way of characterising valleys, I am not merely objecting to a terminology which compares valley form with human life, and is used in a confused and inexact way; I am interpreting the nature of the phenomenon in an essentially different way. It should be readily acknowledged that the detailed study of stages of development has explained many phenomena and enriched our knowledge; but stages of development are given far too much prominence at the expense of different kinds of development; valley forms, like the features of the land surface as a whole, are characterised in too one-sided a fashion. I disagree with this viewpoint because I see it as a schema, and a pretty tedious one at that, rather than an animated rendering of reality.

When an outstanding research worker perpetrates such an error and many others enthusiastically assent to it, or at least tacitly adopt it, we need to enquire into the psychological grounds for this. I believe the answer lies in the scientific methodology used: in the preference for deduction and in the particular form this takes. Deduction will always tend to accentuate a single sequence of cause and effect, and the simplest possible one at that, while others are neglected, especially those not amenable to a deductive approach. In Davis's case the one-sidedness lies in his preference for geometrical interpretation and his neglect of the diversity of natural phenomena, a diversity which can be understood only by the penetrating observation of detail, a diversity which must be treated 'physiologically', to use Passarge's expression. Davis's approach rests on two simple geometrical constructions: that valley bottoms gradually but progressively assume a regular curve which becomes flatter and flatter; that valleyside slopes becomes gradually flatter and at the same time lower until they are nearly horizontal. The gradient of valley sides, like that of their bottoms, is presented as a function of time, or at least of the stage of development reached as a result of the rate at which processes are effective and the length of time during which they have operated. The diversity of rock properties together with the different character of the modifying processes, factors which would be given equal weight in an inductive study, are far removed from this deductive geometrical scheme, and are not allowed for.

It is unnecessary to examine W. Penck's approach in a similarly detailed fashion. His scheme likewise depends on deduction. Rate of uplift, for Penck the critical factor, plays a similar role to that of age in Davis's system. Most of the objections on which Davis's theory is to be rejected can, with little modification, be levelled against Penck's as well.

CANYONS AND OTHER VALLEY FORMS

Valley form is not a simple phenomenon but a complex one comprised of the gradient of the valley bottom, its breadth, the presence or absence of valley terraces and the form of valley sides. These three components stand in a certain but variable (*nicht zwingend*) causal relationship with each other. Each obeys different laws. We should therefore scarcely expect to achieve a unified, all-embracing classification of valleys. We must be content with establishing types, and even then we cannot get very far.

Many valleys seem to be entirely or largely the work of incision by running water, modified to a lesser degree by other forces. Many Alpine ravines are of this type and we can therefore speak of a ravine type. But naturally not all valleys which are called ravines in a local dialect or by tourists are in fact such; conversely, valleys which are not usually called ravines in fact may be of this type. They appear to be associated in origin with former glaciation; perhaps they are often formed beneath glaciers by meltwater. If this is not so the hardness and smoothness of the old glacier bed may have hindered the attack of destructive forces on it. In either case glaciers seem to have acted as the godfathers of ravines. It is debatable whether true ravines are found in unglaciated areas; but many of the valleys of karst regions, once parts of cave systems, can be so called.

Canyons have sometimes been treated as ravines; but this is wrong. Even the most magnificent and typical canyons do not have such a purely erosional character. The Spanish word 'cañon',[25] meaning tube or cannon, has been applied to the deeply-incised, steep-sided valleys of the North and South American Cordilleras, and has been adopted by geomorphology as a generic term. Attempts have recently been made to change its use: on the one hand, it has been given a wider meaning and applied to all V-shaped valleys thought to be young, such as those of the Rhine and Moselle; on the other hand, its use has been limited to the deep narrow valleys of tabular or plateau landscapes.[26] It is just as inexpedient to widen the meaning of the term canyon as it is to use the term fjord of any inlet in a rocky coastline; it is an unjustified departure from the word's original meaning, and throws aside a distinction characteristic of the term. When used in the limited sense we have spoken of, it was with the Canyon of the Colorado in mind; the fact that numerous valleys to which we have to apply the term also occur in tilted strata and in massive rocks (*Massengesteine*) was forgotten. The canyons of tablelands do in fact display many distinctive features; they are a sub-type.

All canyons display three principal characteristics: the river occupies the whole of the valley bottom, river-bed and valley bottom coincide, and the sides of the valley are very steep and hardly dissected. Even the upper edges of the valley are close together, when broad terraces are not intercalated.[27] But the sides of canyons are usually stepped not vertical; moreover in their partial dissection they manifest the work of weathering and degradation, and in this respect they are essentially different from ravines. The narrowness of their valley floor points to the fact that the river is still incising, with the result that lateral erosion has as yet been unable to widen the valley bottom. But we cannot explain the steepness and slight dissection of their walls solely or even chiefly as a result of their 'youth'; it is due to the feebleness of weathering and degradation. As Dutton recognised correctly, true canyons are limited to land with a dry climate, and it is to this climate that they owe their form.[28]

Even in moist climates many valleys are reminiscent of canyons in the narrowness of their bottoms and the steepness and undissected nature of their sides. But their canyon character is less well defined; it is therefore better to speak of them only as canyon-like valleys, but without trying to be pedantic about it. The likeness has been suggested for the valleys and chasms (*Grunde*) of Saxonian–Bohemian Switzerland;* but it is also true of many valleys of the Swabian–Franconian Alb, of the Tarn gorge in the Causse region, and many others. In areas of moist climate these canyon-like valleys are associated with pure quartz sandstone or pure limestone; the permeability of the rock has a similar effect to dryness of climate; we have seen that climatic and lithological aridity can in general simulate one another. Rainfall cannot erode in a dry climate because it is altogether lacking, on permeable or dry rock because it percolates into the rock; where it emerges it undermines to form vertical rock faces. Rock clefts are only of indirect importance inasmuch as they promote the penetration of water and the development of vertical rock faces; in clayey sandstones and other rocks of low permeability they do not have this effect. Canyon-like valleys are generally incised into horizontally-bedded strata and therefore tend to have terraced sides; but they can also be

Age and Form of Valleys

associated with steeply dipping limestone strata, like the Doubs, or with any other permeable rock.

All other valleys in moist climates have more inclined and more dissected sides. The description V-shaped has established itself for these valleys to distinguish them from glacially modified U-shaped valleys; Passarge styles them notch valleys (*Kerbtäler*). In the detail of their form they vary a lot because of both structure and peculiarities of climate. In any one type of rock the steepness and form of valley sides are alike according to the rock, steeper or flatter, more or less dissected. Where the rock is horizontally bedded, terraces and slopes alternate in the cross-profile. Steeply dipping and folded beds give rise to variations in the longitudinal profile, hard rocks protruding as bars, weaker rocks forming more gentle slopes. In a uniformly moist climate with a continuous plant cover gullies and runnels seldom form; in any case soil creep would rapidly smooth them out again; slopes are uniform and smooth and cut only by individual small valleys. In dry regions where rain falls periodically and there is a more sporadic plant cover, the smallest area is permeated and intensively sculptured by gulches and runnels.

In general all these valleys have a narrow valley bottom, one largely limited to the river bed, from which the sides of the valley rise directly or nearly so. But in other cases lateral erosion at a time of equilibrium, or even deposition, has formed a broad valley bottom, either sporadically or along the entire course of a river (cp. p. 24). Passarge therefore calls them flat-floored valleys (*Sohlentäler*).[29] The sides of the valley then lie farther apart. But this does not materially influence their form; valleys with broad bottoms can also have steep walls and look somewhat like canyons.

When valley terraces have been formed during periods of erosional stillstand the valley has what might be called a tiered construction. But this manifests itself only in the width of the valley and not in the form of its sides; these will have receded farther at their top but are no more markedly bevelled there than at their bottom. The structural style is the same at all levels, provided that altitudinal differences of climate or rock variation do not cause weathering and degradation to vary and thus the form of the valley as well.

Besides genuine valleys, in other words valleys formed by running water, the sides of which are the outcome of bevelling and dissection by denudation, there are others which though of the same origin are incompletely developed or have been modified by other forces; these we can call pseudo-valleys. They are of two principal kinds.

Wadis are the valley form of deserts. Their winding nature and the marked contrast between slip-off and undercut slope indicate that they have not been formed by wind but by running water, whether this is the run-off of the more violent downpours which fall from time to time in even the most extreme desert, or a relic of a moister climate in the past. Since they are streamless in the periods between downpours a valley bottom cannot fully develop; but the wind can deflate all the sand from one locality and accumulate it elsewhere, with the result that the valley bottom consists of alternating shallow hollows and gentle swells. The sides of wadis are as steep as those of canyons.

The other principal form of pseudo-valley is the glacial trough or U-shaped valley. Their twistings are diminished, in plan they are fairly straight, and at the same time wide and open, so that one can look right along them. In longitudinal section they show a step-like alternation of flatter and steeper stretches, often even basins which have been subsequently filled in with gravel, and rock bars through which the river now cuts in ravines. In cross profile they have a trough or U-shaped form, a form which can scarcely be ascribed to lateral talus cones alone. Tributary valleys often end at a considerable height above the bottom of the main valley and, to use Gilbert's expression, 'hang'.* There has been a lot of argument over the origin of this form of valley. Many scientists, including Ramsey and Tyndall, have attributed them to glaciers alone, and A. Penck, on the basis of his detailed study of the German Alps (1882), would accept very significant glacial erosion. But others, and in particular the Nestor of Swiss geologists, Heim, will not hear of this, and reject especially glacial overdeepening; they ascribe only *roches mountonnées* and striated rocks to the work of glaciers.* Opposition to glacial erosion seems to be growing. In no instance were such valleys originally formed by glaciers; they were established by rivers as ordinary valleys. Later and perhaps repeatedly, they were modified by glaciers, and then finally reshaped again by running water. We are still uncertain of the form the valleys had before glaciation, in particular whether the stepped structure characteristic of most glacial valleys, and their associated feature of hanging tributary valleys, are a relic of pre-glacial times, and only somewhat modified by glaciers, or whether they did not exist at all before glacial erosion took place. Probably only a comprehensive comparison with unglaciated high mountains in warmer climates will bring us nearer to a solution, since we can in this way substitute one unknown factor for the two of today. Most modern researchers into the glaciation of mountains (*Alpenforscher*) are much too inclined to limit their view to the Alps.

CHAPTER V
Alignment and Arrangement of Valleys

When we turn to drainage patterns and drainage systems, the first question that confronts us concerns the relationship of valleys to the alignment of mountains; this leads to the second and more difficult problem of their relationship to internal build, to the dip and strike of strata, and to the course of tectonic lines: folds, faults and the direction of jointing. The two problems are as one when the direction of strike is the same as that of the mountains; they are distinct when the strata strike across the alignment of mountains, especially when this is determined by major fault-lines which truncate the folds. A great deal of attention has long been paid to these questions, and while it was assumed that in unfolded areas valleys were independent of tectonic lines, their relationship to faults and clefts has recently gained greater prominence. Deecke* in particular has strongly emphasised this, relating not only the overall course of valleys but even their individual twistings to faults and fissures. There is no need for me to discuss this relationship in detail here; I need only emphasise (cp. p. 14) that this is a description of fact and says nothing about the origin of valleys; as far as origin is concerned there is still room for differences of opinion. An important point concerning terminology follows from this. Observations on the relationships of valleys to internal build must not be confused with postulations as to their origin, and the one term used for both. This has happened recently, especially when Davis's school uses terms like consequent, subsequent, obsequent and the like; the two must be kept clearly apart.

CONSONANT, INCONSONANT, SURVIVING AND SUBSEQUENT VALLEYS

According to our conception of the nature of erosion we see a river as having flowed initially above its present valley on the original surface, and gradually incised itself. We can at once propose that a valley has had an erosional origin only when the original surface slopes uniformly in the same direction as the valley. The river can make its way over smaller surface irregularities after dammed-up water has overflowed. But major interruptions in the uniformity of the gradient make erosion difficult to understand. All valleys which accord with mountain structure can be said to have a tectonic origin or be termed consonant (*rechtsinnige*) or concordant valleys.[30] They depend upon internal build in various ways: their alignment can be determined by the slope of a surface or by the course of a line of dislocation, be it a downfold or a fault; we can therefore distinguish between slope and dislocation valleys.

In fact, valley formation is more often determined by faults or other tectonic lines than one usually assumes. Certainly this explanation was once too readily invoked. It is true that a fissure or fault in the structure must first be proved before it can be used to explain a valley. And it is true that it is usually a question of a fault having an indirect influence, when it juxtaposes rocks of different resistance at the surface. In areas where the internal build is still not known with exactness, or where it is very complicated, we must exercise restraint.

On the other hand, we should not be too ready to reject this explanation. It is most unlikely that the Neckar valley of Swabia or the major longitudinal valleys of the Alps or Andes originated through subsequent erosion because the rock was less resistant. Here it is certainly a case of tectonic lines; A. Penck has recently ascribed them to large-scale downwarping in a major folding of the mountains.

But many valleys do not accord with internal build; they are inconsonant (*widersinnig*) or discordant.

The discordance is often no more than apparent, and with respect to the present surface, not the tectonic surface; discordance disappears when we have a clearer understanding of build. As early as 1865, Gümbel referred to the Altmühl as an example of this.* Flowing from a low hilly area it breaks through the escarpment of the Swabian–Franconian Alb. But originally the gently rising limestone surface which forms the Alb extended beyond its present limit, and the Altmühl was established on it as a slope valley; the river became discordant only with the degradation of the land in its upper reaches. According to A. Penck a similar situation exists in the Danube gorge, through the Swabian Alb and in many other breach valleys.*

A river often cuts through folds or indeed through strata tilted in the opposite direction to its flow; but then it is a case of being on an oldland (*rumpfgebirge*) which is sometimes still covered by horizontally disposed strata. The tectonic surface on which the river established its course is unrelated to the folds and the dip of the strata; it is determined by the dip of the remnant surface or its overlying horizontally bedded strata, and the river's course accords with this. Ramsey and others long ago deduced this process from the remnant landscapes (*rumpflandschaften*) of the British Isles; Richthofen termed such valleys 'epigenetic' (Powell used 'superimposed').* A covering of thick glacial gravels also seems to be able to give rise to such an apparent discordance between river courses and internal build as, for example, with the incision of the Danube into the southern edge of the Alb and the Bohemian Massif.

To explain the discordance between mountain structure and river courses, as we shall later see in more detail, earlier English geologists had already postulated a more or less arched surface passing through the ridges and summits on which the river pattern was established.* This device has recently come into great favour with Davis's school. The model is easily set up; but the large-scale nature of processes planing off a mountain and re-elevating the levelled mass must be kept in view. We must seek simpler explanations before we resort to such powerful means. In the Alps and in mountains of similar construction we must examine whether the development of major transverse valleys is not to be related to the nappes which have been thrust forward over the now-existing folds: whether, that is to say, it is not a question of only apparent discordance between transverse valleys and internal build in this case too.

With other valleys their inconsonance or discordance, their being at variance with internal build, is not apparent but real. Such inconsonance is most striking in breach, or corridor, valleys piercing right through mountains. It was with reference to such valleys that attempts to explain inconsonance were first made; theories to explain inconsonant valleys in general have been set up only gradually. They follow two directions.

In the one explanation, forecast by Bischof and Lyell, and used by Powell for the Green River breach of the Uintah Mountains, by Medlicott for the Siwalik forelands of the Himalayas and later, independently, by Tietze for the mountains of Persia, many rivers are older than the mountains and, cutting into them during upwarping, have maintained their old course intact.* Powell terms such rivers and valleys 'antecedent', and the term has passed into the German literature too. Penck described them by the rather ugly word 'preceding' (*vorgehertäler*), Ostreich as 'persisting' (*bestandige*), and I at one time as 'original' (*ursprünglich*), but perhaps it is better to speak of surviving (*überlebenden*) or pre-existing (*preexistierenden*) rivers and valleys. This explanation is of greatest importance in explaining breach valleys, but it is by no means valid for all such valleys. In many of those which have been interpreted as 'surviving' breaches, such as the major transverse valleys of the Himalayas, subsequent origin is more probable. Be that as it may, 'surviving' valleys, like the breach of the Rhine through the Rhenish Massif, are exceptions. In general, internal build constrained the river to adopt its initial course.

The other explanation, the one suggested by Beete Jukes (1862), Rütimeyer, Heim, Gilbert, and used by Löwl (1882)* to explain breach valleys, sees these valleys as being younger than the mountains, not formed at the same time as the mountains were built and part and parcel of them, but developed only subsequently by a variety of processes. Such valleys can be termed subsequent (*nächtraglich*)[31] if American usage is preferred.[32]

Subsequent valleys may be formed in three ways. In an area as yet undissected, new valleys can be formed. This is the case on plateaux where rivers have the gradient necessary for erosion only at the margins of the plateaux, whilst away from these margins rivers flow sluggishly or their waters even sink into the ground. In such a situation erosion is mediate. But this is usually confined to tablelands and other plateaux, while in fold mountains, fragmented block regions and volcanic landscapes, it can usually begin at once, and is immediate. Secondly, the subterranean valleys of cavernous limestone areas can become sub-aerial ones by roof collapse. I am inclined to believe, from observations in the Swiss Jura, that nowadays we often underestimate the importance of this process. Thirdly, rivers which because of the steeper gradient of their valleys or their greater volume of water can incise their course more rapidly than others can successfully encroach upon the area of these other rivers and tap them either from one side or at their source. Davis calls this latter method 'beheading', Penck 'cutting off the roots'.

Davis and his pupils have traced in detail the formation of subsequent valleys and the competition between valleys. In doing so they have explained the origin of many valleys. But they have at the same time been too one-sided in their approach. Differences of rock hardness is for them the main cause of unequal intensity of erosion. As they see it, when rocks of different hardness outcrop at the surface —they often invoke the idea mentioned above of mountains being surmounted by a remnant—surface rivers cut valleys in the weaker rocks, and these rapidly extend headwards to tap neighbouring valleys. Since rock zones generally follow the long axis of mountains, this should give rise in the main to subsequent longitudinal valleys, as Beete Jukes first asserted. Subsequent transverse valleys are added to these, and termed by Davis, and by younger German geomorphologists in pidgin German, 'obsequent' or 'resequent' according to their direction.

Here I shall disregard the fact that the terms 'hard' and 'weak' are too inexact; and I will not comment on the fact that rock hardness is often not observed but only deduced from the drainage pattern—a circular argument. But I must protest against the hasty and one-sided resort to rock type. Subsequent valleys can have other causes, and the major longitudinal valleys of the Alps can scarcely be explained without reference to tectonic causes.

When the two sides of a mountain are of different steepness, the different intensity of erosion on the two sides brings about a displacement of the watershed. The headward advance of valleys on the steep western slopes of the Black Forest towards its gentler eastern slopes is a familiar example. The influence of a marked contrast in the wetness of two sides to which Krümmel* referred can hardly be questioned. It is most obvious when the river has cut back on a rainy slope through the crest and penetrated into a dry interior zone. This may be seen very clearly at the breach of the La Paz river[33] through the Eastern Cordillera of Bolivia; the transverse valleys of the Himalayas may have the same cause, having cut back more or less deeply into and beyond their steep and wet southern flanks.[34] Sometimes transverse valleys seem to have been initiated by glaciers flowing over the cols of lateral crestlines lowering the cols; the river has followed in the path of the glacier.

ORIGINS OF VALLEY ALIGNMENT

Thus there is a great variety of standard causes for valley alignment and arrangement. Any attempts at explanation must consider all the possibilities. Since of its very nature the deductive approach assumes a particular cause and deduces reality from this, it can easily lead to one-sidedness. Nevertheless deduction shows only that an explanation is possible, not that it is correct. It does not show that the development deduced corresponds in fact to the actual development; subordination to an invented scheme is not authentic explanation. Scientific study must initially be inductive and analytical and start from the observed facts; then beginning with the simplest, it must examine all possible explanations before it can sustain a single explanatory principle.

In investigating valley alignment various facts are to be noted. Firstly, we must examine its relationship to internal build and to the original tectonic form and surface which existed, or is assumed to have existed, before degradation commenced. This is sometimes very simple, at other times very difficult, because when the destruction of mountains is far advanced, the reconstructed tectonic surface must remain doubtful; nevertheless, it is necessary to examine this relationship if we are to reach a more certain judgement as to whether a valley is consonant and concordant, that is to say is tectonically determined, or inconsonant and discordant, and at variance with structure. Secondly, we must examine the age of valleys. By studying old valley floors which have survived as terraces, it can be established whether the valley already existed before the most recent uplift, and whether or not the river of that time was roughly equal in size to the river of today; that is one of the most important indices for distinguishing surviving (antecedent) valleys from subsequent valleys. Valley form, the existence of valley divides and the smallness of rivers in relation to their valleys can lead us to conclude that the river has shifted its course. The similarity or contrast in the alignment of neighbouring rivers indicates complementary or contradictory histories. To decide between the various causes of subsequent valley formation we must find out whether the valley coincides with the steeper slope of the tectonic surface, with greater wetness, or with less resistant rocks. An explanation based upon rock 'weakness' is uncertain until this has been directly ascertained. A piece of research work which does not bother with all these types of evidence is unreliable. Davis's own studies rest upon painstaking examination of the facts; but many of his disciples have let themselves be seduced and, omitting exact observation and inductive study, have given us deductive fantasies in place of scientific facts. We cannot content ourserves with a study of individual valleys; it is necessary to study their nature as a whole, the drainage pattern and drainage system; geography must always take account of the whole landscape. Statements to this effect were made at a pretty early stage; but from the difficulty of the subject we did not quickly get beyond such tentative statements and one-sided assessments. Two periods, each with a different approach, can be distinguished. The first considered only, or for preference, how drainage systems were related to internal build; for example, it recognised the regular alternation of longitudinal and transverse valleys in simple fold mountains as well as the arrangement of longitudinal valleys to form a through-valley. The second, on the other hand, studied subsequent modification, how one valley contrasted with another, and their conflict at the watershed, an approach which to a certain extent discards relationship to internal build; at least it tends in another direction. In doing just this Davis and his pupils have undoubtedly rendered great service; for the consistent working out of an explanatory principle always serves to bring more order and clarity into a mass of phenomena. Unfortunately, they have used the deductive method in too one-sided a fashion, and neglected to compare their conclusions with the facts.

Davis maintains that the subsequent adaptation of rivers to rock brings about a particular and recurring valley alignment, for example the longitudinal valley of weak rocks; as a result the drainage pattern takes on a distinctive inconsonant or discordant character unlike the original, tectonically determined and consonant or concordant pattern. In this way the age of the landscape is important for its appearance since the formation of subsequent valleys requires time, and they become more and more recognisable as time goes by. While at the start all valleys are tectonically determined, or are surviving (antecedent) ones, subsequent valleys later become more and more common and often dominant. The influence of the tectonic surface becomes progressively less important than that of the working power of rivers. We must examine in the light of the facts how far this principle is valid. Davis also maintains that the density (texture) of the drainage pattern increases with age. This is not usually the case, except for plateaux landscapes; in mountain chains it may even be, as Richter thinks,* that smaller valleys can be swallowed up by larger ones.

And for the drainage pattern too, the way in which valleys are formed is more important than their age. Just as the original drainage pattern differs with different mountain structures and in fact whenever conditions vary, so subsequent modification is equally varied. Even at the most advanced stage of destruction different structural types will show different drainage patterns, because their rock arrangement obeys different laws. This will be much more true of landscapes at an intermediate stage of modification, in maturity to use Davis's expression. One of the most important aims of geographical-geomorphological research is to be able to appreciate typical differences between drainage patterns according to their different conditions of developmental history, of internal build, and of climate. It has, at least in the present state of our knowledge, the advantage over the 'stage of development' approach of directly linking landforms with tectonics and with climate. And it can in this way help to explain in some degree the distribution of landforms—a matter always of the greatest importance to geographers—while we can still say virtually nothing about the geographical distribution of landforms grouped by age (*Altersstufen*).

Thus the arrangement of valleys and the origin of drainage patterns is an exceedingly complicated problem, one which can only very rarely be completely solved owing to the disturbed nature of tectonic development in most areas, and the commonness of changes in climate.

CHAPTER VI
Benchlands, Remnant Surfaces, and Other Planations*

Among the most impressive traits of many landscapes are the extensive surfaces referred to as levels or planations. Although they undulate and are frequently broken by isolated hills rather than being perfectly even, they are nevertheless flat enough to merit their title. Sometimes they lie almost at sea level. But they are generally at greater elevation, and are then often, though certainly not always, dissected by drainage systems of greater or lesser density. They are composed of older rocks, not younger deposits laid down by rivers, glaciers or the wind. They rarely, if ever, coincide exactly with bedding planes thus enabling us to directly associate their origin with the deposition of strata; more often than not they truncate the bedding planes obliquely, and in Supan's apt expression are 'truncating surfaces (*schnittflächen*). They must therefore be features of destruction and degradation; the remnants of superposed strata which have frequently survived upon them also suggest this. But such truncating surfaces need not be remnant surfaces (*rumpflächen*); the latter term is properly used only for those cases in which the whole landscape has been more or less levelled. Nor can we immediately refer to them by the roughly synonymous term of the Davisian school, peneplain (*fastebene*), since after Davis's definition this term also implies levelling of the whole landscape until rivers attain the profile of equilibrium, even if it does not do so etymologically. Mixing the two terms has introduced much unnecessary confusion into discussions.

How has Nature brought about such even or undulating surfaces, since direct observation more often shows that degradation results in uneven surfaces and only deposition gives rise to flats? This is a difficult problem, and it is understandable that it is nowadays the one that geomorphologists concern themselves with more than any other, and is the most debated in the geographical study of landforms. And this is the more so when the problem is coupled with the claim that lands have been subjected to frequently recurring powerful uplifts.

BENCHLANDS

The great benchlands (*landterrassen*),[35] widespread in just those three countries where scientific research has been fostered most, were the first instances of such levels to be studied. In South-East England, North-East France and Lorraine, and in South-West Germany, scarplands (*Stufenlandschaften*), in which continuous escarpments (*Landstufen*) alternate with broad levels or benchlands, are well developed; those of South-West Germany must serve as a paradigm. Not that the whole landscape is planed off; in fact it is characteristically irregular, mainly consisting of scarps, in front of which rise more or less isolated hills, regularly alternating with levels. It is therefore wrong from the start to confuse benchlands with remnant surfaces. Theoretically, a remnant surface (or peneplain) above the upper edge of scarps can be postulated—we shall discuss this later; but in their present form scarplands are not remnant surfaces.

As a glance at any geological profile will show, escarpments are not fault scarps (*Bruchstufen*); they are degradational features and, indeed, the work of very considerable degradation. Entire geological formations—in South Germany for example, all the Triassic above the Bunter Sandstone, and the whole of the Jurassic—have been consumed; such escarpments have been called cuestas (*schichtstufen*). Certainly, it is obvious at first sight that they are related to the rock. Their upper part consists of chalk, limestone, or pure sandstone. These rocks extend back from the scarp to form the elevated surface, and only at some distance back do they give way to clays, marls and other weak rocks, which in turn comprise the lower part of the succeeding escarpment. On the strength of this, Swabian geologists claimed at an early date that scarp formation depended on rock differences within horizontally bedded strata. But less careful attention was paid to the fact that benchlands do not coincide with the same stratum over their whole area; their surface slopes less than the dip of the strata. In other words, backwards from the scarp edges, benchlands transgress beds higher and higher in the geological succession. They are not structural (*schichtflächen*) but truncating (*schnittflächen*) surfaces.[36] When this fact was more clearly recognised later on it was used by many people to argue that scarp and terrace formation did not depend on rock; and even today this myopic view still figures prominently in the literature. But the fact that benchlands transgress strata does not prove them independent of rocks. By calculating how much of the surface area was formed by different groups of strata, Wagner* showed that the levels are largely restricted to the weaker, less resistant groups. The crest of the scarp is invariably associated with resistant rock. That benchlands depend upon rock cannot be gainsaid, and must be our starting point in any explanation.

By being content to simply assert that escarpments and benchlands are related to rock and making no attempt to explain how the degradation took place, and to relate their origin to valley incision, we have left a great gap in our knowledge. In 1882 Tietze first pointed this out forcefully.* We need to be clear about this. What force can remove material, and work in such a way that degradation produces level surfaces? There is no trace of evidence of marine action—an explanation to which earlier geologists always had first resort. Wind, which Walther readily invokes later under the influence of his work in deserts, is out of the question in moist climates; and we cannot always resort at once to complete changes of climate. It is more likely that cuestas have to be attributed to a former moister climate rather than to desert conditions. An ice cap cannot produce levels (*ebenen*). The effective cause must therefore be the land's surface water. But how can this form levels? These considerations led Powell and Dutton* to the idea that degradation had taken place in a state of erosional equilibrium. This idea was founded upon their work in the area of the Colorado Canyon where at some considerable height, smaller terraces excepted, the so-called Esplanade suddenly broadens out, and only well back from its edge again rises in cliffs. Following their idea I came to the conclusion when studying Saxon Switzerland* that benchlands must take their form from present-day or former valley bottoms surviving as valley terraces. But even at that time I also emphasised the difficulties of this explanation. Rock influence seemed to be of secondary importance; certainly in Saxon Switzerland rock differences are not so impressive, and at that time we did not yet have an accurate geological survey. At roughly the same time this line of thought was the beginning of Davis's peneplain theory,

Benchlands, Remnant Surfaces and Other Planations

a theory which now dominates the outlook of many geomorphologists.

But since then Davis himself (1900)* has shown that the Colorado Canyon Esplanade takes its form from the rock not because it is an old valley bottom. Revisiting the scarplands of Swabia I was myself convinced of the almost obtrusive influence of rock type (*gesteinsbeschaffenheit*), and though I sought it, I could not find any uniform slope of the benchlands towards old valley bottoms. Thereupon I re-examined and changed my views on the levels of Saxon Switzerland, pointing out that they had nothing to do with old valley bottoms.[37] But this did not prevent Rassmus and Staff* interpreting not only the extensive levels but also the many smaller, often only shelf-like terraces in rock faces, and the surfaces of tabular hills (*Steine*) as peneplains in the narrower genetical sense.[38] And despite its absurdity this interpretation still haunts the literature. The more carefully one studies the scarplands of Swabia and Franconia the more convinced one becomes that scarp and terrace formation depends on rock, not old valley floors. Even Gradmann,* who long continued to relate at least the Muschelkalk surface to old valley floors, has abandoned this interpretation. I believe that no-one who knows the country still sees the lowest parts of benchlands (*Fusslinien*) as old valley floors.

Following Gilbert and Richthofen, Philippson and even Sölch prefer to ascribe this surface denudation to the rivers themselves, which, in shifting hither and thither in a state of equilibrium, consume all save a few remnants of higher ground.* But such a view is not supported by observation; moreover, it is theoretically unlikely that rivers could form a valley floor wide enough to explain the levels. Their cause must lie in the lesser processes of degradation; on this, both the majority of those who propound the peneplain theory and their opponents agree. While wind may do this in deserts, in moist climates it must be the work of trickling streamlets and sheet wash. Differences of opinion culminate in the issue whether the level at which the degraded surface lies depends upon present-day or old valley floors or on particular resistant rock strata—in other words, whether such surfaces are the outcome of erosion terraces or denudation terraces being broadened.

Unless every benchland is to be ascribed to a new 'cycle', as some research workers have actually maintained, the peneplain theory requires us to postulate that rivers have flowed along the foot of scarps (Davis's subsequent rivers) and that their several valley bottoms could be brought to one and the same level. But such longitudinal valleys are in fact exceptional; and when they do occur the fact that they are juxtaposed with scarps is either fortuitous, or they have been established subsequently as a result of scarp formation. Extensive benchlands cannot be seen in the field to slope towards rivers; on the contrary they disregard them and sometimes even slope in precisely the opposite direction. Only where the river valley cuts the lowest point of the benchland does the bench fall to the present or old valley floor, and even this is true only when the river still flows on weak rock; if in the meantime it has cut downwards into underlying resistant rock, the benchland's lowest point is still at some altitude above the valley floor. But if benchlands are considered peneplains in the orthodox sense of the word, the fact that they have nothing to do with valley floors is inexplicable; the matter can be taken as settled.[39]

In my paper on the rock forms of Saxon Switzerland I explained rock platforms (*Felsplatten*) and levels as the result of the gradual retreat of rock faces undermined by percolating water emerging above bedding planes or impermeable strata. Then Schmitthenner traced the process further, attributing surface denudation to dell formation.[40] Dells, shallow depressions found on all plateaux, had hitherto attracted little attention, or were taken for 'senile' valleys. They were now put into their rightful place as the agents of one of the most important forms of degradation. They bring about the undulating dissection characteristic of all plateaux formed by degradation. They are associated with impermeable rock, and develop faster or slower according to rock hardness. They do not develop on permeable rock; dell formation more or less ceases when it reaches the surface of such permeable rock. Gushes of water may still excavate a gulch, but will no longer denude the surface. Surface denudation of weak impermeable rocks contrasts with the linear degradation of permeable rocks; this contrast becomes very evident when both types of rock are interbedded in a tabular landscape.

The actual effect of this process depends on the rock composition and the way it is disposed. Where rock composition varies greatly from place to place, many small scarps and terraces develop. Thicker uniform beds form a few high escarpments and extensive benches. In horizontally-bedded rocks only vertical development takes place;

on moderately-tilted rocks escarpments are more likely to lie abreast of one another even if widely separated.

The Swabian-Franconian scarplands, where the strata dip gently south-eastwards and eastwards, can serve as an example. The uppermost weak strata, insofar as they were ever present at all, have been degraded to small remnants. To the south-east the thick strata-complex of the white Jurassic limestone forms the surface; but instead of being a level structural or truncated surface it shows scarp development on a small scale. Canyon-like valleys, in part dry, are incised into it. On its north-western side, the Alb forms a major escarpment with outliers and residuals as a result of the gradual destruction and retreat of the scarp face. From here, deeply-incised valleys cut right into the limestone tableland against the dip of the beds. Less resistant strata lying beneath the limestone form the lower part of the scarp and a part of its foreland. Subsequent development has not been the same everywhere because rock composition varies even within the same bed; but, generally speaking, terraces and scarps succeed one another in the Dogger, the slates or limestones of the Lias, the two or three Keuper sandstones, the Hauptmuschelkalk, the Wellenkalk and the Bunter Sandstone,* until finally in the Black Forest, Odenwald and Spessart the remnant surface of the basement-complex, probably of ancient origin (Permo-Carboniferous), appears as the last bench.[41] Its fall to the lowland plain of the Upper Rhine is a scarp of another kind, a fault scarp probably of more recent origin than the escarpments.

This kind of surface configuration is limited to strata-complexes of low dip. It is true that where strata are inclined more steeply, on the Rigi for example,* the alternation of hard and weak rocks is also evident. But the contrasts are compressed into a limited area, and the surfaces are markedly inclined instead of being slightly sloping levels. Such extensive, markedly-inclined scarps are called 'Hogs' Backs' in the American West; the ridges of the north-west German Uplands also have this appearance in large part. In strata which dip even more steeply, scarp development is limited to valley sides where rows of sharp projections and indentations rise one above another; I have described them where they are developed in a particularly characteristic fashion in the Cordillera of Bogotá.* Levels cannot come about like this; planations on more steeply inclined strata are of another origin.

D

The original drainage must reflect the dip of the strata. When degradation forms benches and scarps, some of the rivers will retain their old alignment and break through the escarpment. Rivers which are in fact consonant or concordant then appear inconsonant or discordant. But at the edge of the escarpment, for example where the Swabian Alb falls away to the north-west, new rivers develop and flow from the escarpment against the dip of the strata; although apparently consonant they are in reality inconsonant (or to use Davis's terminology, obsequent). They cut headwards into the escarpment to varying degree and sometimes tap the upper valleys of the original rivers, so that one can cross a valley watershed from one river basin to another. Davis described this from the Swabian Alb and other regions.[42] But inconsonant rivers can also have originated before the strata were raised and be surviving (antecedent) streams; only after detailed study can we judge how they did in fact originate.

It is often difficult to determine the age of benchlands. They are manifestly younger than the youngest strata forming terraces and scarps, and more recent than the uplifts which brought the strata to their present position, or they probably began to be formed at the same time as these uplifts, for scarp formation begins immediately uplift takes place, but thereafter progresses slowly. Most present-day scarplands are probably rather young, and of Tertiary age at the earliest. In several instances, for example near Alpirsbach in the southern Black Forest, and on the Katzenbuckel in the Odenwald, rock fragments in intrusive volcanic necks indicate that when they were formed they were still overlain by a great thickness of strata which has since disappeared. On the other hand, Oligocene and Pliocene gravels and clays on the surface of benchlands point to their advanced development in the Tertiary period. Their age relative to valleys is generally difficult to determine since the two processes are independent of one another. Levels are formed high above valleys, and thus valley incision need not be younger than the levels. One involuntarily thinks that it should be; and indeed the majority of geomorphologists have accepted that it is. The fact that levels separated by valleys and formed independently of one another lie at the same altitude is due to their common rock composition and arrangement.

To explain scarplands Ramsey thought it necessary to postulate that the whole complex of inclined or mildly arched strata had formerly been truncated (*Kappung*) by a plain of marine abrasion, because only in this way could the weak beds be destroyed. Davis's school follows Ramsey except that for a plain of marine abrasion it substitutes a 'peneplain' of sub-aerial origin. Having done so, the school scarcely bothers to explain what is precisely the problem, how scarps and terraces are formed, but dismisses it with the word 'excavate' (*ausraum*). To postulate such a truncation surface does not make it an observed fact, as has sometimes been naïvely assumed; it is a conceptual aid. Nowhere can these levels be seen. They are in fact unnecessary. Without such a truncating surface either of marine or sub-aerial planation upstanding parts will be somewhat evened-off by degradation; complete evening-off such as earlier planation would promote just does not in fact exist. Neumayr, Penck, Richter,* and recently Gradmann, the latter with particular force,[43] have pointed out that degradation intensifies with increasing altitude. In doing so, one stresses height above sea level with its greater frost weathering, another, manifestly more correctly, mainly greater elevation above the floor of valleys. The greater this is, the deeper rivers incise, the more readily weathering and mass-movement can take place, the faster and greater destruction and degradation will be. Whether we consider, for example, that the inclined platform of Swabia and Franconia was structurally complete when degradation began, or accept the more likely premise that degradation to a certain degree kept pace with gradual uplift, the uppermost groups of strata must have been first and most completely degraded where uplift has been greatest. Thus over the Black Forest and Odenwald only strata lower in the geological succession could survive, and on their western edge only the basement-complex; not until we go farther eastwards in the direction of the strata's dip can we still find the strata of the upper part of the succession.

The actual shape of the ground corresponds with such a conception, arrived at without assuming previous truncation. The Tertiary planation of South-West Germany can be relegated to the realm of fantasy. Schmitthenner's detailed study of Lorraine, and Weber's of Thuringia seem to lead to a similar conclusion, despite the work of Philippi and his followers;* it will presumably be confirmed elsewhere. Thus even in this case there are no grounds for accepting that 'peneplains' occur in scarpland areas.

OTHER PRESENT-DAY PLANATIONS

Benchlands are the planations characteristic of tabular landscapes, that is to say of extensive areas of horizontally-bedded strata of varying resistance; apart from bedded rocks proper they may also develop on extensive basalt cappings. But levels also occur in areas of folding and of massive rocks; and even the levels of tabular landscapes can have an origin other than that already reviewed.

Strandflats are a special class of levels. Varying in width they can be found along many coastlines. Sloping gently seawards they can be traced in one direction sometimes to a considerable height above sea level, in the other to far out beneath the sea. The Scandinavian strandflat, especially that along the west coast of Norway, where its inner edge is sharply demarcated against the backing highlands, or that described by Theobald Fischer* along the west coast of Morocco, are the best known. Although many research workers like Ahlmann* accept a sub-aerial origin for these strandflats on the basis of the peneplain theory, it is nevertheless likely that they were formed by marine abrasion through wave action. Even as zealous a follower of Davis as D. W. Johnson is now returning to this interpretation. Indeed he applies it to levels for which it can hardly be valid. His interpretation differs from that of Richthofen, who founded the theory of marine abrasion, only in that he considers a sinking of the land unnecessary, since he attributes the greatest effectiveness to abrasion when sea level is constant. Naturally, the best evidence for marine abrasion is a covering of marine deposits and the remains of sea animals. But surface irregularities are invariably present and differ in distribution between marine and sub-aerial deposits; this can therefore serve as evidence. Marine formation is readily seen when the rock platform ends abruptly against a cliff face, precisely because the line of the cliff base marks the maximum extent of the sea.

Other rock benches are associated with altitudinal zones of climate. That they could originate in this way was pointed out long ago by Richter.[44] When glacial corries are gradually displaced backwards, their floors can survive and amalgamate with the floor of adjoining corries to form corrie platforms when the dividing walls are degraded. Corries lying on opposite sides of a ridge will get progressively closer, the ridge separating them becomes thinner and thinner until it finally disappears altogether and the corries of both sides unite. Where there was formerly a ridge there is now a mildly-arched plateau which we can term a corrie platform. But when incomplete, corrie platforms are difficult to distinguish from old valley floors and opinions about them often diverge widely. In mountains of Alpine structure high surfaces such as these will be small, since they are part of individual ranges; but they recur on all ridges. On a continuous highland like that of Norway they can be more extensive; but it is doubtful whether in fact such high surfaces as the Norwegian Fjeld originated in this way. Location at the level of the firn line, and glacial action, would serve as evidence of such an origin.

Passarge[45] sees the rock platforms of the Alpine meadow zone above the upper limit of forest as being formed in a similar way; degradation is much more intensive here than in woodland due to the openness of the plant cover. But he has not proved such an origin; and it is unlikely that this is the correct explanation, since degradation in woodland is not as unimportant as he estimated.

I do not deny that there is a tendency for the landscape to be planed towards a base-level of erosion; the question still is whether it has been carried through to completion, sometimes in the face of other tendencies. Sea level will be the general base-level of erosion, and contiguous with this, the bottom of large valleys. But there are also local base-levels of erosion.

In karst regions, the places where rivers percolate into the ground, usually at the lower end of poljes, are such local base-levels. The land around the sink-holes and the floor of the poljes is flattened and levelled on all sides. Of course this does not form a true plain because the different planations always remain separated from one another by surface irregularities and ridges of rock. Since the relief is slight, a cursory glance can give one the impression there are continuous remnant surfaces, the irregularities and ridges being seen as anomalies of no importance or formed later. The single valleys which incise into such high surfaces from their margins are generally, or at least often, formed by the modification of subterranean cavern systems. The low density of rivers and valleys always remain characteristic.

The areas in which rivers dry up, Richthofen's regions of interior drainage, are also areas with a local base-level of erosion. In contrast to what happens in limestone landscapes, here rivers must deposit débris and fill up the hollows. Though the thickness of débris infillings may not be as great as has been believed, and is often only a thin veneer to a rock bench—as, for example, at the foot of the Castilian Cordillera Central, or, according to Passarge, in the Schotts of Algeria—nevertheless distinctive degradational as well as depositional processes must be effective. Certainly the wind cannot alone be responsible for this; rather we must claim that water works in a distinctive way conditioned by climate. This is undoubtedly relevant to a problem to which special attention has been given in recent decades, the origin of inselberg landscapes.

Inselbergs, hills rising sharply (*ohne Fusshalde*) from gently sloping sea-like plains (and therefore reminiscent of an island),[46] were first recognised as a peculiar phenomenon by Bornhardt in East Africa. They were found again and studied more closely by other German workers, Passarge, Jaeger, Thorbecke, Obst, Waibel, Schmitthenner, Maull* and others, especially in various parts of Africa, but also in eastern Asia and Brazil. Yet many were scarcely noticed by Anglo-Indian geologists in India, although it is in this area that they particularly lend themselves to study. As far as is now known in whatever parts of the earth they occur they do so largely (or only?) on old massifs, but are limited to periodically moist, savanna climates, and extend with certain modifications into the temperate zone only in the monsoon climate of eastern Asia. Their formation seems to depend on the marked contrast of seasons: in the dry period, very intensive mechanical weathering occurs, often in the form of rock spalling or desquamation; in the rainy period, the powerful effect of surface run-off occurs, the so-called sheet-flooding (*schichtfluten*).[47] As a result valleys are much broader than in our climate, and stream-less valley-like depressions common, which is why Brandt* called the Río de Janeiro landscape 'valley-less mountains'. Here the planation process used to explain scarplands does in fact take place. Rock plains become more and more extensive, hills shrivel up and become isolated residuals. It is scarcely conceivable that rock platforms should do otherwise than slope towards sea level, broad valley floors or local base-levels of erosion.

THE THEORY OF REMNANT SURFACES

Having studied planation as it occurs today, we can turn to the problem of remnant surfaces.

In 1846 the prominent English geologist Ramsey, studying the

Uplands of South Wales, noted that all the ridges and peaks have approximately the same altitude, and look as if they have been cut out of a more or less level plane which Ramsey considered once formed the Uplands surface. It truncates the steeply-dipping and folded strata and thus has nothing to do with old foldings and dislocations, or indeed with internal build at all. On the contrary, it is at variance with internal build and must be due to a degradational process.[48] This observation and conclusion was the beginning of the theory of remnant surfaces.

Understandably, English research workers tend to lay particular stress on the effect of the sea and give it prominence. Ramsey accordingly explained remnant surfaces as the result of marine degradation, that is to say the planing effect of wave action, and linked it causally with the 'transgression' of superposed strata. Later Richthofen, unaware of Ramsey's theory, made the same observation about the mountains of South China. He too explained the formation of these plateau surfaces which truncate whole mountains as the work of marine 'abrasion',[49] since he considered sub-aerial forces inadequate. He went further than Ramsey inasmuch as he considered the land must have subsided for this to take place. He applied this same interpretation to the mountains of Europe, where the discordance between structure and surface could not have been overlooked, but was unthinkingly taken for granted and dismissed under the general heading of degradation. While Ramsey's remarks went almost unheeded, Richthofen's attracted great attention and recognition. For a number of years research was more or less under the influence of this theory until a new one of sub-aerial degradation and planation took its place alongside the old and suppressed it. But theory swings back and forth like a pendulum, and of late more attention has again been given to marine abrasion. I have already mentioned the theory of D. W. Johnson (the editor of Davis's *Essays*), which returns to the theory of marine origin.

The new theory was first developed in America and can be seen as a continuation and expansion of Powell's and Dutton's theory on the origin of benchlands. Certainly, Neumayr[50] and Penck[51] advocated that sub-aerial degradation could go as far as to form a remnant surface, or a lower level of denudation (*Denudationsniveau*), as Penck expressed it. I extended the theory of the end-state of erosion (*erosionsterminante*) to explain how sub-aerial degradation might reach the stage of planation. But Davis developed it in an all-embracing fashion, initially to explain the planation of the Pennsylvanian Appalachians; he later applied it to many other areas as well.[52] His theory goes beyond Powell and Dutton inasmuch as it is also applied to folded areas and asserts that whole lands undergo planation. Rivers incise gradually to reach the profile of equilibrium, and at the same time valley sides are increasingly bevelled and flattened until they scarcely rise above the horizontal surface. At such time as the flanks of adjoining valleys intersect, the watershed is little above sea level, the whole land has been virtually planed and become a peneplain (*Fastebene*). This interpretation is closely linked with Davis's theory of life-phases. The peneplain, in other words an almost complete planation, coincides with the stage of senility, the end of the cycle; after this a new life and the development of new landforms will be possible only if the land is uplifted. With unflagging enthusiasm he repeatedly proclaimed his theory and gained for it far-reaching recognition. The fact that the French geologist Lapparent accepted it and expounded it at the Berlin Geographical Congress of 1899 was especially significant for its adoption in Europe.* Linked with the cycle theory, that is to say with the theory that the land surface is revitalised, it has until recently been the prevailing doctrine.[53]

Before we can take account of the nature and significance of the great planations of the earth's surface formed by degradation we must adopt a simple and unequivocal terminology. Unfortunately, advocates of the two theories use different expressions. When Richthofen and his followers speak of surfaces being abraded they do not mean abrasion or planing in general, but merely marine abrasion, that is planing by wave action. Davis, on the other hand, introduced the linguistically neutral term peneplain, but coupled with it a genetical significance, namely planation by sub-aerial forces until rivers attain the profile of equilibrium. With such terminology we can speak neither of abrasion surfaces nor peneplains if we want to leave the origin of a degradational surface open. Instead we must say: this surface is an abrasion surface or a peneplain. But science demands that we first seek a descriptive term which designates the phenomenon independently of any theory of origin. Any term founded on a theory is transitory and hinders agreement with the supporters of another theory. A good descriptive expression is the one introduced by Richthofen: 'remnant surface', which implies not only the extensive planation but equally its discordance with internal build and the fact that the tectonic surface has been so destroyed and degraded that it is unrecognisable. Only the remnant (*rumpf*) of the original mountain remains. If the planation is especially striking one can speak of a remnant plain (*Rumpfebene*).[54]

The remnant nature, that is to say discordance between surface and internal build of a degree approaching planation, is most evident in steeply-dipping and folded strata, since the surface is then in manifest discordance with the strata's bedding. For this reason, many attempts have been made to restrict the terms remnant and remnant surface to old fold mountains. But such a limitation is not justified by the nature of worn-down remnants; remnant surfaces have been brought about by sub-aerial or marine degradation removing great masses of material in blocklands, tablelands with horizontal or gently-inclined strata, and volcanic regions also. Here, however, discordance between structure and surface is not so immediately apparent; often it can be established only after detailed study. Benches which end in escarpments must not be confused with remnant surfaces; tablelands become remnant surfaces only after the escarpments too have been more or less planed off. I do not know whether such tabular remnants exist; but there is no theoretical reason to doubt that they do and limit the term 'remnant' to remnant folded areas.

Both theories see the formation of a remnant surface as an event of the greatest significance in the earth's history; but they do so in different ways. Planation by marine abrasion is probably linked with subsidence of the land (or a rise of sea level), allowing the sea to encroach upon the land. But such movements need only be moderate. Even the greatest mountains can be undermined by aggressive wave-action. Though sub-aerial planation does not call for subsidence of the land—nor must there be uplift either—sub-aerial forces certainly need a long time to bring about such planation. Large valleys are incised slowly enough; considerably longer is needed for the smallest valleys to be incised, even more for valley sides to be flattened and the landscape planed to a state of senility. In considering the age of valleys, I have already stressed that this point is taken too lightly, and too little account given to the tremendous length of time needed for valley formation. This is still more true of the planation of whole lands and the formation of remnant surfaces or peneplains. Brückner has the planation of the Swiss folded Jura, their re-uplift and folding, take place in part of the Pliocene, and rejected Machatschek's objection as immaterial.* This means he must ascribe an immeasurably long duration to the Pliocene. Geological periods of such length are not otherwise necessary. Certainly in geology one is accustomed to think of geological periods as long, but a certain moderation should be observed; we should not be too extravagant with time.

Few remnant surfaces display unequivocal evidence as to whether they were formed in one way or the other. This is especially true of fossil remnant surfaces formed in the more remote geological past, and most of the clearly recognisable remnant surfaces are such. Later dislocations, superimposition of deposits, and degradation have generally so obliterated their original layout and configuration that it is now difficult to judge what they were once like. Only with detailed and painstaking study of the age and nature of overlying deposits and all the conditions of the relevant geological period can we arrive at even a likely conclusion as to how such fossil remnant surfaces were formed. But although it must be seen as among the most important geological problems, the origin of such fossil surfaces is of no direct geographical interest, since it has nothing to do with the present landscape. Geography takes these old remnant surfaces for granted.

According to both the marine and sub-aerial theories, planation will come about close to sea level, in one case somewhat below it, in another a little above. When we then find remnant surfaces at great altitudes and dissected as we usually do, both theories require us to assume that the surfaces have been subsequently uplifted and dissected. The major role played by remnant surfaces in modern geomorphology rests precisely on this assumption, and on its being coupled with a recognition of periods or cycles of erosion.

If the land stops sinking or subsidence is offset by uplift, marine abrasion will cease at a particular contour, where the remnant surface will abut sharply against higher ground to form a cliff. The theory of sub-aerial planation, on the contrary, must postulate that the landscape is evenly degraded over its entire surface and can only be combined with a gradual backward increase in the height of the ground. The theory cannot explain level rock platforms at the foot of mountains, the so-called piedmont flats, or even a flight of such surfaces lying one above another, or piedmont stairways, insofar as they are not the result of subsequent tectonic disturbances. At least, in many cases when remnant surfaces have been claimed, too little attention has been given to the fact that as yet we have no plausible explanation of how piedmont surface and piedmont stairways have been formed sub-aerially.

Irregularities in surfaces of marine abrasion will be related largely to differences in rock resistance and therefore rather sporadically distributed; those of surfaces of sub-aerial planation, on the other hand, will always take the form of valley systems and watershed ridges, though they too are adjusted to differences of rock resistance. On the former, rock permeability plays scarcely any part; the latter show differences of degradation as between permeable and impermeable rocks. The composition of overlying deposits will also be different in the two cases; a covering of marine strata probably indicates marine planation, and sub-aerially formed strata, sub-aerial planation. But even here we can be deceived. The evidence is compelling only when it can be established that deposition followed immediately after the remnant surface was formed. A remnant surface of marine origin can later be overlain by continental deposits, a sub-aerial one by marine deposits. Thus in Saxony, the Upper Cretaceous lies on the remnant surface of the crystalline basement complex. Since the Upper Cretaceous succession begins with deposits formed sub-aerially, a sub-aerial origin for the remnant surface has been inferred. But isolated, limited occurrences of marine brown Jurassic strata indicate that the surface is much older, and was once covered by marine strata.

Remnant surfaces of sub-aerial origin must show the characteristics of 'senility', in particular that is, wide valley floors with quite flat sides. Dells, as ubiquitous on remnant surfaces as on benchlands, must not be taken for senile valleys; nor must we forget that the presence of escarpments argues against 'senility' of the landscape.

To be candid, we have to admit that neither of the two hypotheses will enable us to understand fully how remnant surfaces are formed. I can say this more impartially about sub-aerial origin since I was among the first to try to establish it. When I oppose such an origin I am not arguing against an idea new and foreign to me, but criticising myself. I still believe such a planation of the land is possible, but it must be a process of extraordinarily long duration; and I wonder whether enough time has been available, and whether the conception is otherwise so warranted that we can disregard these misgivings. Davis and his pupils founded the peneplain theory solely on a geometrical construction, that both valley sides and valley bottoms are progressively flattened, not on the observation of levelling processes. Despite the objections raised against it, they are much too ready to accept this theoretical construction, with a certain scientific apathy, as an established fact about which we need trouble no more. In doing so they neglect to lay the foundations of their doctrine and one day it will collapse about their heads.

Sölch and, in a more comprehensive treatment, W. Penck have given the theory of sub-aerial planation a somewhat different turn.* In keeping with their views that uplift is slower they would have planation begin at the same time as uplift and keep pace with it, not take place after uplift has finished. In other words, the land does not first become mountainous at all, as internal build might suggest, but remains flat and thus imperceptibly becomes a remnant surface. The remnant surfaces we find in nature are not final-remnants (*Endrümpfe*), as with Davis, but pseudo-remnants (*Trugrümpfe*—Sölch) or primary remnants (*Primärrümpfe*—W. Penck), or at least they can be. A remnant surface is raised so that we now find it on mountain tops, and is gradually being attacked by rivers working inwards from its margins only when mountain building, and this in the form of simple uplift, later speeds up.

Their approach thus aims to obviate the scarcely conceivable magnitude the process must have, if a high mountain is to be degraded to sea level and entirely planed off in a relatively short time only to be re-uplifted. It seeks to make the occurrence of more recent remnant surfaces at great height above sea level, easier to understand, as well as the case where a number of planations lie one above another. The approach certainly overcomes several of the objections I have expressed towards peneplains in Davis's interpretation. But it still leaves planation as a powerful and scarcely imaginable process, and there is as yet no support for the view that it has actually occurred in our climate.

Extensive planations, completely discordant with internal build and therefore remnant surfaces, exist today, but only in association with inselberg landscapes. Thus the question arises whether the fossil remnant surfaces of the geological past, from which indeed solitary hills (monadnocks) nearly always rise, are not also of the same origin. This idea was in fact first suggested by Passarge;* it has continuously gained ground, especially with regard to the widespread Permo-

Carboniferous remnant surface. The strata overlying this surface, the New Red and Bunter Sandstones, indicate a seasonally moist tropical climate, such as is characteristic of present-day inselberg landscapes. Strigel in particular has propounded this view with respect to the remnant surfaces of the Black Forest and Odenwald.[55]

THE OCCURRENCE OF REMNANT SURFACES

If we want to review the occurrence and distribution of remnant surfaces over the earth, we cannot assume that every planation is one. And we must distinguish much more than we have between fossil remnant surfaces, the emergence of which took place in the remote geological past and has now ceased altogether, and recent remnant surfaces which were formed in at least more recent geological time if not the present, and conform with present-day mountain structure.

One of the most noteworthy facts is that the majority of well-preserved remnant surfaces which are of any significance at all for the landscape's appearance today are geologically old, and frequently have been overlaid by more recent marine or sub-aerial deposits, and affected by faults and uplifts since they were formed. Some seem to be of Cambrian age, and many are Devonian and succeeded by the Devonian folding, not only in Northern Europe but also, according to Chudeau,* in the Southern Sahara. The best known remnant surfaces originated shortly after the folding of mid-Carboniferous times and the faulting and volcanism of the Upper Carboniferous and Permian ages, and thus are of Carboniferous and Permian age. This holds not only for those of the German Uplands, the Plateau Centrale of France, South-West England, but also those of the Appalachians, the Urals, the Tienshan, and the northern Sahara, etc. All of them are already part of the internal build (*gehören dem in neren Bau an*), and, from the geographical point of view, completed features; we can leave their explanation to geology. A problem for geography is how they have been freed of the strata which overlay them and modified in their form to such an extent that, according to some research workers, young remnant surfaces have been cut into them, in recent geological time.

Few remnant surfaces of Mesozoic and Lower Tertiary age have been established with any certainty. It remains to be seen whether few were actually formed or if they are difficult to recognise either because of the strata's tabular arrangement or because they have been involved in recent folding.

The question of greatest interest from a geographical standpoint is whether, and to what extent, remnant surfaces have been formed in geologically modern times, since the more recent foldings, faultings and volcanism which determine present-day structure. Apart from strandflats such were rarely spoken of until recently; under Davis's influence they became a fashion in American, German and French geomorphology. Many research workers consider it an incontrovertible fact that they exist, even that whole series of them exist. To prove that such modern remnant surfaces or peneplains exist is central to their studies, and almost amounts to 'good form'. With the greatest facility they decide that complete planation followed by renewed uplift and erosion has taken place. In fact one of them has even gone so far as to see this approach as the very beginning of scientific geomorphology.[56]

The first to do so with reference to the German Uplands, Philippi,* tried to establish a relatively young, but nevertheless pre-Oligocene remnant surface (or as he says less precisely: a land surface) in South-East Thuringia. He claimed that it crosses directly from the old schistose mountains to the Muschelkalk, generally missing the Bunter Sandstone, and that it planed off the dislocations dating from the beginning of Cretaceous times.[57] Without examination, often without looking for evidence, other research workers recognised this planation in different German landscapes: Maull, and likewise Krebs,* laid a young peneplain across the Spessart and the Muschelkalk landscapes drained by the River Main; W. Penck and Schrepfer* plane off the southern Black Forest including the Baar, in the Pliocene; and Braun has constructed a 'Germanic remnant plain' of Miocene age across the entire German Uplands.[58] Briquet, too, laid a remnant surface across North France.[59] In the oldlands of the Frankenwald, the Erzgebirge, and the Rhenish Massif, these young remnant surfaces are seen as contiguous with older ones, but distinguishable from them. In the areas where strata arranged in tabular fashion have become scarplands they are seen partly as coincidental with the benchlands, partly as passing over the scarp crests, so that at least the lower parts of the scarps manifest subsequent excavation. Even towering young mountain chains are said to have been levelled more or less completely to sea level after folding, and then re-uplifted, and to owe their mountainous nature to this re-uplift and its consequent renewal of valley formation. Braun propounded such a view for the Appennines, Martonne for the Transylvanian Alps, Brückner for the Swiss Jura, Staff in extreme form for the Alps, Ostreich even more extremely for the Himalayas. Philippson had a peneplain crossing the East-European lowland, and Bailey-Willis* even theorised one through the whole of Asia.[60]

One is immediately struck by the fact that unlike old remnant surfaces, these young 'peneplains' so freely presented to geomorphology, insofar as they actually are remnant surfaces or peneplains and not benchlands or other local planations, have never survived intact, not even rarely, and therefore cannot be directly observed. On the contrary, they are inferred from other phenomena thought to be inexplicable without them; in other words, they are hypothetical constructions. That recent peneplains lie concealed beneath recent alluvial lowland plains (Philippson) is also a speculation which is without doubt only exceptionally true. One suspected this from the start. Recent remnant surfaces should really be better preserved than older ones subjected to destruction for a much longer period. I cannot find a statement of any sort in the literature which explains this contradiction. It could be said that the land was immediately and intensively re-uplifted after recent planation, whereas a quiescent period in which the surfaces were covered by terrestrial and marine deposits followed older planations. But there is no factual support for such a difference between the two.

We cannot examine here all the remnant surfaces claimed, as to whether they are necessary and do in fact exist; for this we would need to carry out detailed investigations on the spot.[61] Philippi's pre-Oligocene land surface in Thuringia, which has become the model for so many other peneplains, rests more upon fundamental study than most others do; nevertheless it is not a fact actually observed but a hypothetical construction and we must examine its validity. What one sees is a typical scarpland; only its lowest member, the lower Bunter Sandstone, rises to form a surface above the remnant surface of the Frankenwald (folded schists) at Ronneburg, in just the same way as it does in the Black Forest or Odenwald. The Muschelkalk surface again comes to roughly the same height as the remnant surface of the schistose mountains (*Schiefergebirge*). But between them the Bunter Sandstone forms a depression inconsistent with the peneplain and for which no satisfactory explanation is given. It is dismissed as an 'excavation'. Other assertions also provoke criticism. A new exact investigation is needed. Meanwhile, let us keep an open mind, especially since Hans Weber* established that on the Gossel Plateau, as well as in the Tambach region, scarplands can be explained without having to assume former planation.

CRITERIA FOR THE RECONSTRUCTION OF REMNANT SURFACES

Here we can do no more than critically examine the criteria and evidence on which remnant surfaces may be reconstructed.

Levelled surfaces which are discordant with internal build and transgress major faults are the most direct and, on the whole, most trustworthy basis on which to do this. Such surfaces are usually horizontal or slightly inclined; but markedly inclined slopes of smooth surface and extensive development have also been considered, for example by Brückner* in the Swiss Jura, parts of a former remnant surface which has been refolded or tilted. But smooth surfaces can also be formed locally, and need not be seen as fragments of more widespread planations. They can be benchlands, corrie platforms, the local planations of karst regions or areas of interior drainage, etc. They must be numerous and widespread and occur on various substrata to be seen as parts of a more general planation; they must also show that they are related to present-day or old valley floors. To found such a phenomenal process as the planation and re-uplift of the Himalayas on a single elevated surface seems inadmissible to me.

Deposits of river gravels or loam spread widely across surfaces show that their sub-surface was completely formed by the time deposition took place. Except in regions of interior drainage, such depositions will usually be associated with lowland plains and will therefore indicate that uplift has subsequently occurred. But they show evidence that a remnant surface (peneplain) was formed before or at the time of deposition only when they are widespread and rest on neither a bench or other local planation, nor a fossil remnant surface.

A third criterion on which it is claimed remnant surfaces may be reconstructed is the accordance of summits, the existence of a summit level (A. Penck's *Gipfelflur*).[62] Certainly, when one looks about from

Benchlands, Remnant Surfaces and Other Planations

peaks the uniform height of surrounding summits is often truly impressive; understandably it is those who know the High Alps who stress it. To a certain degree, this is an optical illusion; the vertical dimension is so much smaller than the horizontal that differences of height become inappreciable; they are much more apparent from the valleys. But actual uniformity is by no means so astounding. Direct degradation will also cause most summits to be roughly the same height; a remnant surface is unnecessary. In general, the greater the altitude the more intensive the weathering, and summits therefore tend to become equal. But it is much more important, as Richter* has noted, that valley sides cannot exceed a particular angle, and that all eminences rising higher than this normal slope and the divides along which the sides of two adjoining valleys intersect must fall prey to destruction. This brings about a certain conformity in the height of summits, the more so the farther advanced the degradation. There is no need to attribute summit levels to the hypothetical destruction of a remnant surface.

Discordance between internal build and surface form is a fourth starting point from which we can reconstruct old remnant surfaces. This discordance can amount to complete inversion, so that where there are tectonic archings or horsts, valleys or other depressions occur, and where tectonic troughs or trenches, ridges. When, as early as the 1860's, English geologists turned their attention to this problem in studying the Weald, they were at a loss to know what to do except postulate the formation of a surface of marine abrasion in the normal course of degradation, to explain how the weaker strata of the anticlinal cores had been attacked by sub-aerial degradational forces. Many modern geomorphologists also maintain that we cannot explain synclinal ridges and anticlinal valleys, or how tectonic horsts have been developed to form depressions and tectonic trenches, unless we postulate a 'peneplain'; where such discordant features occur they see them as support for doing so. Such evidence plays a major role in the work of Briquet and Philippi, and was one of the grounds on which Staff* would have a peneplain through the Alpine summits. And it has even been argued, as we have seen, that for a cuesta landscape such as that of Swabia and Franconia to be formed, the scarp crests must have been joined to form a plain of degradation. But in keeping with the logical rules of scientific research an assumption that complicates the process as this does can be made only if simpler assumptions prove inadequate. That is not the case here; why is a remnant surface necessary and direct erosion insufficient to bring about relief inversion? As soon as valleys are sufficiently incised for their bottoms to be lower than the floor of tectonic troughs or trenches, as can easily happen when mountains are uplifted in their entirety, the intervening mountain mass is seized upon by erosion and denudation, and, according to the resistance of its rock, more or less degraded. As Schmitthenner* has shown, the floors of troughs or trenches can become mountain ridges, the anticlines and horsts, depressions. Intermediate planation is unnecessary.

Reliance is also placed on drainage pattern, especially on the prevalence of valleys which disregard structural folds and faults, or as Staff puts it, are 'indifferent' to them. Such a drainage pattern can only have been developed on a uniformly sloping surface. This being so, the predominance of major transverse valleys which break right through mountain ranges cannot be directly explained by open folding. In this connection Rütimeyer and Heim* long ago recognised the contrast between the valley patterns of the Alps and those of the Swiss Jura. But this contrast has been well enough explained by the way the folds are tightly packed, in which case troughs are not reflected in the relief. And only when, following the more recent tectonic interpretation, we replace tightly-packed folds by nappes inclined outwards, can we adequately explain the prevalence of transverse valleys which are concordant to the original tectonic surface, but discordant and epigenetic in relation to the fold structure developed in the nappes and now exposed at the surface.

River meanders are another basis on which remnant surfaces have been reconstructed. Single bends and windings can develop just as well without them. Moreover, a river can assume a systematic layout of regular twistings along its whole course only when it creeps along sluggishly, and the curves which result constantly increase in amplitude. The presence of a wide valley floor, however, is enough for this to happen; and such a floor can be entrenched into uplands and need not be a remnant surface. And even on theoretical grounds the approach is questionable. It could also be argued that the meanders first developed during incision in response to differences of rock type and arrangement; only a detailed investigation can decide whether this is so or not.

Finally it has been argued that the composition of deposits at the foot of mountains is evidence for or against remnant surfaces: when these are of fine texture no other mountains can have existed where those of today now stand. Braun* has called this the mountain margin method. The reasoning is valid and we must examine the composition of such deposits in any given instance. Thus, Penck* used the coarse rubble present in all late Tertiary deposits at the foot of the Alps to do precisely the opposite, and refute Staff's* claim that the Alps had been planed.

I will not deny that young remnant surfaces can exist. But the many remnant surfaces which crop up in the literature are not, as Philippson calls them, 'incontrovertible observed facts' (*Grundzüge* II, 2, p. 330).* In fact most of them are inferred from uncertain evidence. The origin of landforms, to explain which remnant surfaces have been adduced, can generally be explained in a simpler way; and there is even direct evidence to hand, in particular obvious scarp formation, against many recent so-called planations.

Moreover, the process[63] according to either Davis's 'peneplain' theory or W. Penck's interpretation is inherently improbable. For an entire mountain to be planed off almost to sea level and a cycle to reach the stage of senility is in itself a phenomenal process which if it is ever fully realised, must involve hundreds of thousands or perhaps millions of years. It can hardly be compressed into so geologically short a period as part of the Miocene or Pliocene. And for all that it would be only one half of the process. A peneplain thus formed is then supposed to be re-uplifted and so drastically destroyed that virtually none of it remains, and it can be detected only from indirect clues. There must be weighty reasons for us to make such a revaluation of all our standards. Deduced possibilities are not enough; factual evidence must be provided. The painstaking analysis of landforms must show that simpler assumptions miscarry. Most works of the modern school of geomorphology are lacking in both requirements. Once again we see that to apply a deductive schema to Nature is not the right way to learn her secrets; only painstaking inductive research can do this.

Our discussion thus ends with a large question mark. But since I wrote the article on remnant surfaces and pseudo-remnant surfaces[64] which ends with only doubts and denials, the picture has altered somewhat. We can explain the origin of benchlands by particular processes without any reference to 'peneplains'; and a way to explain remnant surfaces is beginning to appear, though in fact it explains their formation only in particular climatic conditions. But the setting-up of peneplains deductively, and untenable speculations that entire mountains have been planed off in the course of a cycle of erosion, re-uplifted and re-destroyed, must once again be condemned.

CHAPTER VII
Structural Plan and Structural Style of Mountains

THE STRUCTURAL PLAN

To the naïve person, summits, ridges and plateaux, indeed all protuberant landforms, are the features actually shaped, and valleys and other excavations are merely spaces between these, not features in their own right. At first geomorphology also adopted this view. By studying the origin of mountains more closely we came to conceive of them as tectonic features largely the result of uplift, that is, in the way they appear to us in nature. Valleys were seen as open fissures formed during uplift. Such an approach is a necessary transitional stage and should not be despised. We first had to realise that surface features were related to internal build before we could recognise how they deviated from this by being modified sub-aerially. These modifications have proved themselves greater and greater as study has progressed. We have learnt to appreciate how powerful erosion and all forms of degradation can be; it has become more and more evident that the actual surface differs from the tectonic surface. Only when they are recent and still active, and as yet little altered by gullying, do volcanoes have the appearance they owe to their eruption. Very quickly indeed the loose scoriae, ashes and tuff are worn down, the more resistant lava streams and volcano necks left standing, so that we cannot describe them as more than volcanic ruins or skeletons. The basalt and trachyte hills once seen as 'homogeneous volcanoes' are simply ruined volcanoes, in fact in large part only the plugs of old volcanoes which have themselves entirely disappeared. The ridges formed by the arched folding of some mountain chains, the horsts of some block-faulted regions, contrast with the valley-like depressions formed by basins and trenches, but such cases are rather exceptional. We are confronted just as often by the opposite, by inversion, where synclines form ridges and deep valleys are incised into anticlines. It is not only a case of rivers incising valleys; degradation gives rise to broad ridges between the valleys. And we must often or even generally think of mountains as having a particular tectonic form; in other words, we must not think of them as they once in fact were, but as they would still be if degradation had not taken place, towering above all the present crests and summits and fallen prey to destruction.

The primary fact in the sculpturing of mountains is the formation of valleys by the erosion of flowing water. Certainly weathering and degradation sets in even on the original tectonic surface; but the surface then available to their attack is relatively small. Erosion speeds them up, and in forming valley sides exposes new surfaces to their action. Weathering and denudation can then work most powerfully and even shape independent major landforms. They combine to bring about mass-movement, the importance of which was formerly considerably underestimated. But in their direction and spatial arrangement, weathering and denudation continue to depend on the spatial arrangement of valleys. Hence the structural plan (*Bauplan*) of mountains and highlands is determined by their stream and valley pattern (*see* Chapter V); these provide us with the key to understanding mountain masses. For example, individual Alpine masses and their different groundplans (*Grundrissform*), described by A. Penck as ribbed (*fiederformige*) and tiered (*stockformige*), have been distinguished with the aid of valley alignment and density (*Abstande*).

Sometimes a landscape's character is entirely dominated by valleys; the sides of adjoining valleys intersect in linear ridges or crestlines which are no more than boundaries separating adjoining valleys. No other independent landforms stand in contrast to the valleys. Davis sees such a landscape as being at a particular stage in its development, the stage of 'maturity'. But that is only half the truth: many landscapes are in this state from the start, in others it never occurs. It is not primarily a question of stage of development but one of different kinds of surface sculpturing. Volcanoes and indeed most mountain chains, the larger tectonic levels excepted, are well-nigh completely dissected from the start. In contrast, oldlands (*Rumpfgebirge*) and tablelands will remain incompletely dissected for a long time, perhaps forever. In such areas adjoining valleys do not abut directly against one another to form crestlines; upland masses with level or rolling surfaces are left between them. Above them rise isolated peaks, mountain ridges and rock walls, the landscape's upper storey as it were. Tablelands formed largely of moderately resistant and impermeable strata, especially clay, and therefore subject to surface runoff, rapidly lose their original tabular form and become generally dissected valley landscapes which are, however, speedily levelled and lowered by denudative processes. But on permeable limestone and sandstone, and on oldlands of resistant rock, fragments of tablelands and plateaux survive for a long time to manifest a distinctive relief and degradation independent of the valleys.

The difference in altitude between crestlines and valley bottoms, that is the intensity of dissection, or relief energy (following Partsch*), depends initially on the amount of tectonic uplift, using this word to include all uplifting processes whether secular uplift, folding or faulting. The greater the uplift the greater the power of water to erode, the faster rivers incise and approach the profile of equilibrium, while interfluve ridges are degraded at a much slower pace. In advanced age, valley deepening becomes slow while interfluves are degraded at the same pace as before and therefore faster than valleys deepen, with the result that relief is gradually eliminated. So the fact that the Pyrenees are less dissected than the Alps is to be attributed to the somewhat greater age of the former.

It is evident in many mountains that summits and eminences as a whole are associated with hard or, more accurately, with resistant rock, saddles and hollows with weak or less resistant rock. I need only recall the Pfahl of the Bavarian Forest, or the quartzite ridges of the Hunsrück and Taunus. Also the fact that basalt and other volcanic rocks rise above their surroundings is patently due to their hardness. Davis termed such hills of hard rock rising above surrounding weaker rocks 'monadnocks'. Closer study has shown that many, perhaps most summits, those of true mountains as well as tablelands, are formed in rock identical with their degraded surroundings. Their survival has nothing to do with rock hardness. This is true even of Mt. Monadnock, from which the type gets its name; much more often the majority of summits rise where they are farthest removed from the onslaught of destructive forces. As the knowledge that this was so, knowledge familiar to German geomorphology for decades,[65] dawned on the modern school as well, Spethmann called such summits 'watershed residuals'.*

If summits and saddles depend partly on their siting relative to the contours of lowest altitude (*Tiefenlinien*), and partly on rock hardness, the question arises: how are these two factors related? A gradual change seems to take place. The layout of valleys is at first determined by the tectonic surface; it has practically nothing to do with rock resistance or hardness. Summits and saddles are sited in relation to the drainage pattern not to rock resistance. But rock resistance influences subsequent modification, and as time passes, summits and saddles are adapted to it. If rocks vary greatly in their resistance, as for example in volcanoes between the core or lava streams and the incoherent ejectamenta, or in stratified rock between hard permeable limestone or sandstone and weak impermeable clay, adjustment to the rock soon takes place; if differences are less pronounced, progress is slower. But adjustment does not usually appear until landscapes are old; it is most impressive on oldlands where degradation has reached some kind of end-stage. The quartzite ridges of the Hunsrück and Taunus are part of such an oldland. Moreover, relief inversion (*vide supra*) results from an adjustment to rock resistance and can come about directly; it does not call for an intermediate planation, as has been thought.

THE STRUCTURAL STYLE

Besides their groundplan and elevation which together make up their structural plan, mountains have a structural style; this, together with the structural plan, constitutes their architecture. Under the term structural style, a term first used in fact by Richter,* I refer to the way in which mountains develop particular features, characteristic assemblages of landforms or, one can say, a physiognomy.

As with valleys, Davis sought to express the overall character of a landscape by characterising its age. At first sight this aim has something to recommend it; it is understandable that it has met with much approval. But regrettably the concept must be destroyed. Apart from the fact that such an aim fails to consider minor landforms, the development of which largely determines the landscape's structural style, it is wrong to deal with major landforms *en masse*. The state of maturity deduced on the one hand from the drainage pattern and its relationship to mountain build, on the other from the state of valley bottoms, are two different things. Furthermore, the most prominent characteristic, the nature of valley sides, has virtually nothing to do with age.

The extent to which the original or tectonic surface of the landscape survives depends on the intensity of mountain building and on rock disposition and make-up more than on age. These, together with climate, determine the landscape's structural style or physiognomy.

The intensity of mountain buildings, expressed in relief energy, affords a primary means of distinguishing structural styles. Larger rivers at least will usually incise rapidly, with the result that valleys are deeper and summits rise higher above them the higher the mountains are as a whole. High mountains and uplands are to be distinguished only on the basis of their differences in relief energy. We should not introduce here a distinction such as that between mountain chains and oldlands, between, for example, the Alps and the German Uplands. Mountain chains can be of moderate height, oldlands of great height and relief energy. Moreover, not all forms of high mountains can be compared to the Alps. Tropical high mountains have knife-edge divides and peaks only insofar as they reached up into the region of firn during the Glacial Epoch. A second distinction is that between true mountain forms and those of plateaux and tablelands; ridges are generally characteristic of the former, elevated level surfaces of the latter. And thirdly, on a smaller scale, differences of rock type and the way it is disposed are significant, the most impressive being the distinction between permeable and impermeable, coherent and incoherent rock. Every type of rock can be said to have its particular assemblage of landforms for any given kind of rock disposition and climate. Tablelands of horizontally-bedded strata show contrasting forms at different heights; on fold structures such contrasts are juxtaposed; and on homogeneous massive rock no differences at all are apparent over small areas.

But like minor landforms, the entire assemblage of landforms, the structural style, varies not only with the type of rock and its disposition but also with climate, and not only with the climate of the present but with that of the past too. Many research workers, Passarge* in particular, believe that only past climates had any effect in the wooded landscapes of temperate and tropical zones since these researchers hold that weathering and denudation in wooded areas are imperceptible, and landforms correspondingly of long standing—

tenacious of life, to use Sölch's expression. Even in landscapes which owe their structural style to the work of running water, individual landforms vary according to the climate. But the differences of landform wrought by earlier glaciation or by an area being subjected to desert conditions are greater still. Let us begin by comparing two adjoining German Uplands. The Odenwald is a fluvial landscape, a valley landscape. It has been shown that the vestiges of glaciation claimed by Hessian geologists are not so; they are at most the effects of frost action under periglacial conditions. But the Black Forest, constructed after the same tectonic pattern, rises to a greater altitude above sea level and this resulted not only in greater relief energy but in its highest parts being permanently covered by snow, causing small glaciers to descend into the valleys during the Glacial Epoch. Corries, the glacial origin of which was first recognised by Partsch in 1882 in the Riesengebirge,* are a phenomenon typical of the highest parts of these mountains, and in the southern part of the Black Forest moraines, striated rocks and the overall form of valleys point to the former existence of valley glaciers, but which did not descend to very low levels. In their structure the mountains of the British Isles are, like those of Germany, oldlands; but they are different in physiognomy. The more frequent occurrence of corries and glaciated valleys, the more complete replacement and suppression of fluvial (fluviatile) landforms by glacial ones, gives them a generally wilder appearance. In this respect they resemble the Alps, which otherwise they are quite unlike in internal build and therefore in structural plan. In the Alps, elevated surfaces are less frequent and ridges are the major landform dominating the structural plan; they owe their configuration entirely to corries, which have given them their characteristic knife-edge divides. Glacially-modified trough-like valleys extend as far as the foot of the mountains. Similar features, which can be collectively described as Alpine, are to be found in other mountains of the temperate zone; they invariably indicate former glaciation. But in tropical mountains of equal height, for example the Cordillera of Bogotá, there are no glacial landforms at all, or they are restricted to the uppermost altitudinal zones.

We have shown already that in deserts and semi-deserts, in other words in dry or 'arid' regions, minor landforms are different from those of moist or 'humid' climates. In keeping with the different kind of chemical rock decomposition and the absence of surface run-off,

Structural Plan and Structural Style of Mountains

rock forms are not rounded off but are angular and craggy. Rock pillars (*Felspfeiler*) of the most varied forms, including extreme mushroom shapes, accompany lattice-like piercing of rock faces and peculiar caverns. Major landforms are also different; we cannot yet explain the full significance of this. The fact that extensive areas lack any integrated system of valleys can be explained by their lack of rainfall or at least its extreme paucity; but impressive dry valleys, or *wadis*, and large rock amphitheatres or circular, basin-like valleys (*Kesseltaler*), are to be found in many areas. There are three conflicting views about desert valleys. The view that they are hollows deflated by the wind, and not true valleys at all, is untenable in the face of their well-developed valley form. The ephemeral streams of today, ceasing to flow only a short period after they have begun, are scarcely able to form such long, markedly twisted and ramified valleys. Such valleys must indeed have been formed in a period of more plentiful rainfall and greater run-off, and are a case in point of the climatic development of landscapes (*see* Chapter IX).

CHAPTER VIII

Dependence of the Land Surface on Internal Build

Geomorphologists began by seeing the present form of the earth's surface as virtually a direct expression of its internal build. Only gradually did we recognise the major contribution of erosion and destruction, and understand the difference between the actual surface and the tectonic surface. But as a result geomorphology remained in close touch with both geotectonics and geology, remained aware that the reshaping of the surface is always related to internal build. It is only recently that Davis's school has sought to free geography from geology and replace the petrological description of rocks and a geological chronology by a specifically geomorphological approach. His school considers even folds and faults of secondary importance to widespread uplift and subsidence, phenomena which can only be inferred from surface form. Rühl[66] unequivocally advocates the complete separation of geomorphology and geology not only as far as their aims are concerned, a justified corrective to the tendency to include too much within geography we have already stressed, but in their research methods as well. Such a change in approach would unquestionably be of great significance not only to the way work was divided between the two disciplines, but also for the content and soundness of their results. We must therefore be clear as to how far the land's surface features depend on internal build, and to what extent the geomorphological approach goes hand in hand with the geological and must be based upon it, or is independent of it.

When explanatory geomorphology began, the influence of rock type on surface features was sometimes given first place; we find this in Bernhard Cottas's book *Deutschland's Boden,** a work of great importance in its day and still stimulating. He attributes the variation of surface features almost exclusively to variation in the nature of rocks. Similar approaches may be found in many of the older geological books; we sometimes still find them today among geologists, and geographers originally trained as geologists. But research has progressed beyond this stage in two respects. Surface features are the result of rock weathering and denudation; but these vary with climate and therefore produce different features from identical types of rock in different climates; only within a particular climate can we attribute distinctive surface features, or rather a tendency for distinctive surface features to be formed, to one type of rock. Furthermore, the influence of rock type is less than that of the overall nature of folds, blocks and the tectonic surface. It is these which determine whether degradation or disposition predominates, and the way erosion progresses; rock type plays only a secondary role in guiding subsequent valley formation, in the intensity of degradation and in the development of minor landforms.

GEOMORPHOLOGY AND ROCK FORMATION

To this extent German geomorphology has always been in accord with more recent American or Americanised geomorphology. But now the two are diverging. German workers adopt the geological approach to the study and classification of rocks; the American school substitutes a distinctively geomorphological approach. In the works of modern geomorphologists we rarely find the geological names of rocks other than limestone, or they are only mentioned in passing. Instead they speak of weak and hard rocks. Weak rocks favour erosion and denudation, the development of valleys and the lowering of slopes, hard rocks hinder them. But these terms weak or hard are not applied as a result of direct observation; rock weakness or hardness is deduced from the formation of valleys and the lowering of slopes. Rühl[67] openly admits this without seeing anything amiss, surely a circular argument of the first order! When geomorphology began in the 1870's the works of the time also unhesitatingly explained valley formation, their narrowing and widening, the steps in their floors and their terraces, by the alteration of harder and weaker rocks. This circular argument was severely criticised by Löwl,* and research workers have since been on their guard against it. Now, under an American banner, it is again making its appearance. Geomorphology has long since realised that rock resistance is only in part a function of its mechanical hardness, and probably depends more on its permeability and solubility and the way the rock disintegrates. But that was only German geomorphology and did not merit attention. Since it is now proclaimed by Americans as well it is seen as a new revelation.[68]

Each of the rock properties influencing surface features can be studied individually. But this does not entitle us to dispense with the normal geological understanding and naming of rocks, which embraces as far as possible the different rock properties. Of course many types of rocks, for example sandstone and granite, must be divided into sub-types, according to particle size and the material which cements them. If we proceed critically the normal geological interpretation of rocks is an irreplaceable foundation for geomorphology.

There is another reason why it is advisable to use the geological classification of rocks. Their individual characteristics, such as hardness or permeability, must be accepted as given facts by geomorphology. We cannot explain the geographical distribution of individual characteristics, and an approach aiming to do so thereby misses the truly geographical issue. Nor can we as yet completely explain the distribution of rock types; but since this can be done in principle from their origin, every step forward in our knowledge of this also brings us nearer to understanding their distribution.

Geography has a different attitude towards geological formations, that is to say towards the study of geological age, than it has to rock type, a difference reflecting the different aims of geography and geology. Geology, as an historical science, in fact as the history of the earth, must give priority to age. Its primary aim is to construct a chronology. Until he has determined the age of a stratum, a geologist cannot incorporate his observations of its petrological make-up and its fossil content into the earth's history. Geography, on the other hand, is indifferent as to the geological age of rocks and strata as such; rocks and strata are of significance to the geographer only as the material of which the landscape is built, the properties of which depend not on age but on composition. It was thought at one time that rocks of any one age were of a particular type; Humboldt could still group together the red sandstone of South America and the Bunter Sandstone of Germany. Although this view may still be encountered it can be taken as disproved; we now know that the composition of rocks depends on the various kinds of rock forming processes and on subsequent modifications. A geological formation has a particular rock composition only within a limited area and

Dependence of the Land Surface on Internal Build

narrow span of time, and when the conditions under which it was formed were uniform. As a rule reference to geological formations only distracts our attention and unnecessarily burdens the memory. The idea that we could directly reconstruct the distribution of land and water in a particular geological period from the geological map has also had to be given up since we learned that many deposits are formed sub-aerially and realised the measure of degradation. Small-scale geological maps are therefore of very little use to geography, and the smaller the scale the more generalised must be the various horizons and facies. The fact that small-scale geological sketch maps are so frequently found in geographical texts is generally a sign of inadequate attention to method.

We must also consider how surface features are related to internal build or tectonics—a term that can be used in either a narrow or broad sense.

In the more limited and once customary sense, and still the meaning preferred by some people, the term refers only to the way rock is disposed, that is to say the direction of the strata's strike and dip, its faulting, and the rock arrangement thus brought about. This is 'structure' in Davis's geomorphology. The space relationships of rocks of different resistance depend on this, and the origin of subsequent longitudinal valleys, the development of scarps and terraces, and variations in valley width, can be seen as its consequence.

But geomorphology must go beyond this narrow meaning of internal build. The way the rock is disposed is now only part of the picture. For a complete picture we must also know the character of the rock itself, as well as the uplifts and subsidence of warpings which affect whole areas, the great extent and significance of which has come to be more and more recognised. Internal build is the complete expression of the entire formative history, made up of the processes of degradation and aggradation, and dislocations of every kind.[69] It is in this sense that we can speak of types of internal build, of the Rhine Massif as a remnant blockland formed of folded schists and similar rocks, the Swiss Jura as simple fold mountains built largely of limestone, the Alps as mountains formed by much more intensive folding and powerful overthrusting, and perhaps later uplifted as a whole. Major independent volcanic features are also such types, for there is no reason why we should exclude volcanism from the term tectonics.

Certainly internal build cannot be directly observed since it has been more or less destroyed by the processes which modify the surface. More often than not it can be only hypothetically reconstructed. But that does not alter the fact that it was once a reality, still more that it would be a reality had surface forces not worked at its destruction from the start. We can term the surface of this tectonic feature, whether actual or ideal, the tectonic surface. In Chapter V we saw how important this is in understanding the alignment and arrangement of valleys; following on from the result of our discussion there, we can say that it is impossible to understand the actual surface unless the tectonic surface has been correctly appraised. This is where geography leans most heavily on geology. Although tectonics can in theory be considered a part of geography, and every scientific text in regional geography must give us a characterisation of a country's tectonics, research in this field is largely the concern of geology; geography must stand on geology's shoulders.

Yet Davis's school also tries to free itself from geology by adopting a different approach to mountain building. It does not see the folding, overthrusting and faulting manifested in the way rocks are disposed as having shaped the present tectonic form and surface. The landforms to which they gave rise are seen as having been destroyed and levelled long ago, and are now considered only as the cause of the way rocks are arranged. The height of mountains and highlands, so important for the relief of today, is seen rather as the effect of widespread major uplifts and subsidence or warping. Mountains such as the Appennines, the Swiss Jura and the Alps, which have hitherto been considered young fold mountains, are grouped morphologically with the German Uplands. But such an approach robs the word 'build' (*Bau*) of its broader sense and limits it again to its former narrow meaning; an understanding of major tectonic processes and the mechanism of earth movements becomes well-nigh superfluous. Davis sees his explanation of mountain forms as the result of the gradual dissection of uplifted peneplains rather than of a complex history of mountain building, as an extraordinarily important step forward in geography.[70] Supan was more inclined to regard it as the new theory's declaration of bankruptcy.

Many valleys undoubtedly show elevated old valley floors, the altitude and dissection of which must indicate a general uplift of the land. And it seems likely that widespread planation has sometimes

Dependence of the Land Surface on Internal Build

taken place; but nowhere has it been adequately proved that folded highlands have been planed down and then re-uplifted to great heights, or even refolded, entirely within the upper Tertiary period. Yet this has often been asserted. Valley terraces, with the help of which we can trace subsequent mountain uplift, are relatively insignificant phenomena, and of secondary importance to the valleys themselves. They show that the valleys and divides already existed when uplift took place. In other words, the dissection and sculpturing of mountains does not depend on uplift but on their original build. In fact, even a cursory comparison of different structural types: of simple and complex fold mountains (*Deckengebirge*), of block mountains, remnant platforms and tablelands, shows that their relief differs in both major and minor respects.

We must also concern ourselves with the age of tectonic features. Tectonic age is a measure of when the modification and sculpturing now continuing first began. It is at the same time a measure of the age of the present surface configuration, or its morphological age.

It is unusual for tectonic movements to be rapid and simple; it is therefore generally impossible to determine their age precisely. Sometimes, and especially in inland sea areas, two successive processes combine to bring about the present build; folding first, followed by fragmentation through fracturing, uplift and subsidence. Since subsequent uplift, subsidence or warping of whole mountain masses has repeatedly recurred, the age of such movements is important for geomorphology.

A CONCEPT OF AGE

In many localities we can discern traces of several major periods of mountain building each of them separated by long intervals of time; the German Uplands were extensively folded in the middle Carboniferous period, and underwent faulting and block movement from the mid-Tertiary into the Quaternary. Even in the Alpine region, still older mountain building has probably taken place, the traces of which have been more or less obliterated by very much more recent mountain building. Geologically, these events are equally important whether older or more recent, since geology is the history of the earth. They are interesting as historical events; indeed the older ones may be more important than the more recent because they set the stage and pointed the way for the newer. When the geologist speaks of the mountain building of Central Europe he is thinking especially of old folding; in his monumental work *The Face of the Earth*, Suess* always emphasises old foldings more than younger ones.

Geography takes up another viewpoint. Old foldings can now be seen only in the way strata are arranged, in the structure, while the tectonic form and surface of a mass as a whole mountain depend on more recent folding or faulting. This is true not only of young fold mountains but also of block mountains. We see the Black Forest, the Odenwald, the Rhine Massif and the Harz, primarily as block mountains, thanks to their movement in Tertiary and Quaternary times, and only secondarily as fragments of old fold mountains; these areas owe their overall dissection to their block form, and only their detail to their folded nature.

In Davis's school the concept of age has taken on a special form. The age of landforms is measured not against the geological time scale based upon the evolution of the plant and animal world, but on the degree to which they have been modified, the character of landforms, the physiognomy of the landscape. But since their degree of modification depends not only on how long has elapsed but also on rock resistance, landscape age is, as we have seen, not a purely chronological term but one of degree of development. It is paradoxical but true that Davis's school, in which 'stage' is the third word of the trilogy structure, process and stage, in fact altogether neglects age. For geography, the science of the earth's surface as it now is, age is certainly of secondary importance. But it is not unimportant. If we are to understand the laws governing the construction of the earth's surface we must know how features are related to one another in time. Moreover, age is important in understanding all the modifying processes. Only when placed against the geological time scale can we assess the rate at which modification takes place and how it differs with rock type and climate. Climate and the plant and animal world have changed even in recent geological time, and these changes have exerted great influence on the surface modification of the earth's crust. We must therefore know how mountain building and changes in climate are chronologically related.

Thus we find ourselves thrown back upon the geological approach to internal build if we want to explain the earth's surface features. The student of geomorphology may have looked forward to giving geology a rest; he must again take up the hammer and immerse himself in knowledge of rock types and rock formations. Geography must not become geology; nor should the geographical description of a landscape—and here I am in full agreement with Davis—consist of a narrative of geological history, as is too often the case. But our approach to internal build must remain geological. The idea that geomorphological knowledge can be gained without the help of geology, by skirting around it one might say, is absurd. Geology remains the indispensable aid to geographical geomorphology.

CHAPTER IX

The Development of the Land Surface

THE NATURE OF DEVELOPMENT

In a geographical study of the development of the land surface we must take as our starting point the formation of its internal build. It is true that Davis and others before him have regarded the most recent emergence of the land from beneath sea level as the primary state, and that in many cases this amounts to the same thing since major folding appears to take place in submerged geosynclines. But we do not yet know whether this is always so, and whether it is also true of block movements and general uplift and subsidence; it is possible that they can also take place sub-aerially. On the other hand geographers know better than to start by studying the processes of degradation and deposition operating prior to major tectonic upheavals; the only effect of these on the present surface is through the medium of rock composition. The geographical development or, to use for once the imagery so frequently used by Davis, the geographical life, of the present surface of the earth begins when its internal build is constructed. But this does not mean that its construction has to be complete before surface development can begin; in fact this is more likely to take place at the same time, sometimes keeping pace with construction, more often lagging somewhat behind it.

Development of the land surface implies firstly the cumulative effect of processes. Valleys and mountains are not produced in a day, any more than Rome was! On the contrary, according to the uniformitarian viewpoint introduced by Lyell, they are the work of many thousands or millions of years. As time passes the changes wrought by surface forces become greater and greater. One can therefore speak of landforms as having an age, using that word in its literal sense to mean that landforms have developed over a specific length of time. One can also see the successive states of degradation or modification *in toto* as developmental stages, and arrange them into a series of such stages, the whole forming a sequence (*Ablauf*) through which the land surface passes in its development. For the same reason the processes which modify the land, especially those that destroy and degrade it, must be seen as theoretically approaching slowly towards an end-state in which they cease to be effective, either because there is nothing left to destroy or because the forces themselves no longer have the power to attack the land that remains. But only detailed inductive study will tell us in what state of development this should happen and whether it ever actually does so. The controversy now centred on this question comes to a head in the issues discussed in an earlier chapter as to how planations or remnant surfaces have been formed, and how right we are to accept their formation in recent geological time.

But the term development also implies that past conditions which are different from those of today are still effective. Thanks to the ease with which air can be set in motion, weather and climate are related to contemporary causes alone; but because the earth's surface features are more permanent, their causes are rooted in the past. Landforms display not only newly-acquired characteristics but also those inherited from the past; relic landforms (*Vorzeitformen*), to use Passarge's expression, stand alongside contemporary ones. The fact that conditions change in the course of time is of fundamental importance in the study of landforms. But in keeping with the nature and magnitude of these changes, and the landscape physiognomy they produce, different features survive in different localities. Because of their origin, many landforms and landscapes are of the same kind and can be described as harmonious, or better still as homogeneous; but there are others, and perhaps the majority, derived from different periods and of different origins, which are disharmonious or heterogeneous.[71] Indeed, the further geomorphological research penetrates the nature of landscapes the more it is confronted by traces of a different past, even if these traces are sometimes minor or unimpressive phenomena.[72]

Since differences between the past and the present are of two kinds we must distinguish between two series of developmental stages. Even after the major crustal movements to which the lands chiefly owe their internal build have ceased, less intensive endogenic processes such as volcanic eruptions or weak folding and block movement, or what is actually the most common, general warping, secular uplift and subsidence has usually continued to take place. We can describe the whole sequence as the development of internal build or tectonic development. These processes change the level of part of the earth's crust as well as the rock and the way it is arranged; at the same time a new field of work and new material to work upon is presented to the forces which modify the surface. But in addition to tectonic development there is the change in landforms caused by changes of climate, or climatic development. These two kinds of development are quite independent of one another. Certainly, with each major uplift or sudsidence, climate also changes: glaciations have been attributed to this. But in fact changes brought about in this way seem to be subsidiary to those caused by world-wide cosmic or planetary factors. Because of its twofold origin it is impossible for us to find a unitary approach to landscape development.

TECTONIC DEVELOPMENT

The simplest example of tectonic development, and practically the only one considered in Davis's theory, is the progressive but normally discontinuous uplift (its progress may be interrupted or even reversed by periods of quiescence), gentle arching or upwarping of a portion of the earth's crust. Since it is uniform over wide areas or associated with only slight tilting, such movement cannot be discerned from the way the strata are disposed. Because of its gradual nature, it is virtually impossible to recognise such movement directly except along coastlines, where even a slight uplift manifests itself in the withdrawal of the sea. One can argue and argue whether the level of the sea or the land surface has changed; so Suess proposed that the non-committal term 'negative strandline movement' should be used. But since strandlines rise to higher altitudes landwards and have an irregular outline one has no option but to invoke uplift. So the old term 'secular uplift' again appears justified. If the movement is one involving the solid crust of the earth we would expect to see its effect inland as well; the warping of the formerly horizontal strandlines of Lake Bonneville in the Great Basin studied by Gilbert* is probably the same phenomenon as the displacement of the sea shore. The term 'epeirogenic movement' coined by Gilbert to describe it is therefore synonymous with secular uplift and subsidence. But it should not be used in a wider sense to include other kinds of movement. By studying how areas change in appearance away from the coast we might be able to recognise either uplift or subsidence; but such studies are as yet insufficient and too imprecise to allow us to do this. Usually, we can only infer them from the geomorphological features to which they give rise, especially valley terraces; there is therefore always some doubt attached to assertions that they have taken place (*see* Chapter III).

Many mountains display old valley floors lying at a considerable height above the present floor of a valley, and these have long been attributed to uplift. But they have generally been regarded as relatively unimportant compared with folding, and seen only as an interruption in the formation of the valley, of little significance for the appearance of the landscape as a whole. It is only recently that uplift has been declared the really decisive factor for mountain building in general. Hand in hand with this view goes the opinion that periods of erosion are of far-reaching importance for surface configuration.

Davis's cycle theory expressed this approach in its fundamental theoretical form. His theory is not in fact the general theory of landscape development for which it sometimes passes, but is limited to the influence of uplift. It gains far-reaching significance only if we grant that uplift does play a major role, and that intervening periods of stillstand were of long duration. The theory maintains that the history of the land surface begins with its emergence from below sea level and elevation to an appreciable height. Thereupon, the flanks of stream-cut channels are flattened by weathering and mass-movement. Valleys are formed and develop progressively and continuously through different stages of life until they attain an end-state in which their bottoms display quite gentle gradients and their sides slope gently towards these. Their development to this stage was termed a cycle by Davis. After this, uplift may begin anew, whereupon erosion is re-invigorated and a new cycle begins (A. von Böhm* compares it with winding up a clock). Such a development may be repeated again and again. At first, as in his paper of 1894 (*Geographical Essays*, p. 181), Davis distinguished between merely episodic minor interruptions and major ones introducing a new cycle; later both he and his pupils discarded this distinction and considered even the smallest uplift an interruption to erosion and the beginning of a new cycle. The disturbance of 'normal erosion' by volcanic eruptions and changes of climate are seen as complications of the cycle, not as interruptions. The theory does not allow for breaks in the sequence of development caused by new folding or subsidence.

Factually, this theory is not as original as has often been thought. It introduces only a novel form of expression, lays greater emphasis on the significance of uplift and on distinguishing periods of erosion, and considers planation during periods of stillstand more widespread than is commonly believed. But by its consistency and schematic nature it has led geomorphologists to devote more attention to such phenomena than they have hitherto. Distinguishing cycles has become a favourite aim of geomorphological studies. Indeed one can almost say that the setting-up of cycles has become a mania. Their significance must therefore be critically examined.

At any stage erosion and valley formation can be interrupted. This is most readily open to examination when there are unmistakable valley terraces, that is to say elevated valley floors, and when erosion has attained the profile of equilibrium and the state of maturity has set in. Inasmuch as cycles are founded on such conditions no objection can be raised. But in that case, the fact that a new cycle begins is of little consequence for the general character of the landscape.

If uplift is renewed before a broad valley floor has been formed, as is very often the case in small tributary valleys, the interruption in the progress of downcutting does not manifest itself in terraces. Can such an interruption be determined by any other means? Behrmann,* in his study of the geomorphology of the Harz Mountains, took breaks of slope as evidence of this; W. Penck's theories tend in a similar direction. But in fact, breaks in slope have almost nothing to do with the tempo of erosion; they are due to denudative processes.

If, on the other hand, the formation of valleys had already passed through the state of maturity into that of old age, or senility (as Davis describes it), in other words the land had been almost levelled, renewed uplift and erosion will not just form valley terraces; more significantly, the valley will be entrenched into a peneplain. For Davis and his pupils this is a process of the first importance. The formation of peneplains as surfaces of erosional equilibrium, and their subsequent uplift and dissection, are a recurring feature of their discussions. Just as zoogeographers once moved the earth to explain the distribution of a beetle, Davis's school has whole segments of the earth's surface moving up and down, being destroyed and levelled, as though they were stage scenery. But here again, the theory is usually not supported by the facts. Most 'peneplains' are postulated on slender evidence; and such planations as actually do exist may have to be explained in a different way.

The cycle of erosion concept was given special importance in this approach to the study of landscape, because it was thought that it provided a means of correlating what may be called landscape storeys with varying degrees of destruction; destruction at the lowest storey would form only narrow valleys, at an upper one, broad elevated plateaux. On the uppermost storey it would leave only isolated inselbergs. This may, but need not be the case. Such contrasts at different elevations can also result from differences in the resistance of strata. The extensive levels of landscapes such as the scarplands of South-West Germany are benchlands dependent upon differential rock resistance; the small valley terraces are at best the only argument for cycles of erosion. The remnant surfaces of the Black Forest, Odenwald, Spessart, and Vosges are old Permian surfaces that have been exhumed by denudation at an arbitrary height above the bottoms of the valleys. In studying the present form of the land, such surfaces must not be uncritically seen as senile landforms. As far as their origin is concerned they are unrelated to the developmental history of the landforms now surrounding them. Since their formation they have been covered by marine deposits and powerfully dislocated. To postulate cycles of erosion, and to explain differences of landforms at different elevations, are two separate things. In landscapes which have passed through only one cycle, but are made up of rocks varying greatly in their resistance, extensive elevated plateaux can be juxtaposed with narrow valleys. Conversely, in polycyclic landscapes valleys may be of simple shape with only moderately broad valley terraces or even without them altogether.

Since general uplift plays such a major part in Davis's theory—a much greater role than does subsidence—it was necessary to look for a cause to explain this. Rühl thought the answer lay in the theory of isostasy.[73] Degradation of the land destroys the equilibrium of the earth's crust; when destruction reaches a certain point it crosses a threshold of resistance offered by the crust, and the block rises to be dissected anew. This could conceivably happen after large-scale degradation and planation of the whole landscape, and the hypothesis was put forward to meet this case. It would explain why peneplains are only exceptionally found at the height at which they were formed, that is slightly above sea level, and generally at greater altitudes and dissected. But uplift has also been renewed at a stage no more advanced than that in which only valley terraces had been formed; in these cases the preceding degradation cannot have destroyed the earth's crustal equilibrium. Thus the hypothesis is at least inadequate.

In contrast with widespread uplift, arching and upwarping, there has also been extensive subsidence, downfolding and downwarping. Such downward movements are known from coastal areas, and provided that it is the land which has moved and not the sea, they can be traced landwards just as easily as can uplift. They are neglected by the cycle theory and in fact are considered only when subsidence has drowned the lower part of valleys to replace the normal cycle by a marine one. But land which has subsided may still be above sea level. When this happens a valley is displaced relative to the curve of equilibrium. If the river was still incising it can continue to do so only at a decreased tempo and with less energy. If it had more or less attained the curve of equilibrium it will now have to deposit material and build a gravel plain. But this explanation will hardly do for the gravel plains of major Alpine rivers. In their case, Penck's explanation, the infilling of overdeepened glacial valleys, is more likely to be correct; but it certainly appears true that in many other valleys, especially those of tropical mountains, gravel plains have been formed by subsidence or downwarping of the surface. In this way, too, a new cycle would begin, but it has the opposite effect. Renewed uplift will change the gravel plain into a gravel terrace, renewed subsidence may bring the redeposition of gravel. In this way cycles of opposite effect may repeatedly alternate with one another.

Dislocations strictly defined, that is to say folding with overthrusting and block movement, have a different effect from simple uplift or subsidence uniform over wide areas. Recent literature shows a certain tendency to lump the two kinds of movement together under the rubric 'epeirogenic movements'. This is hardly justified. Secular uplift or subsidence affects larger areas, is more periodic in its occurrence, and often seems to follow a pattern of increasing and decreasing intensity. When either involve extensive areas of the earth's crust, and movement takes place in a nearly vertical direction, their effect on rivers and surface configuration is in fact like that of general uplift and subsidence. Remnant surfaces and benchlands will be lifted vertically and retain their form. But it is different when the faults are closely spaced and the crustal blocks more or less tilted, and still more so when it is a case of folding and large-scale overthrusting. These movements are not to be seen as merely a modification of the previous landscape. They entirely transform the internal build and destroy the stream channels and valleys of the older surface; erosion is not just reinvigorated or slowed down, it is shifted into lines of development and must start all over again. Only a few especially powerful rivers will maintain their courses in the face of the upheaval to become surviving (antecedent) rivers in the new landscape. Here the cycle approach breaks down; a new developmental history has been initiated just as when the land sinks beneath sea level and later re-emerges. Geography cannot trace this changing development; it must accept the new internal build as a given fact.

Volcanic eruptions of limited extent produce only an insignificant disarrangement which is usually overcome quite quickly. But those that produce large mountains as well as lava flows and tuff deposits, often of great thickness and extent, give the landscape a new appearance, one scarcely related to the old as folding and block movements do. Despite their youthfulness such landscapes must be considered tectonic surfaces.

CLIMATIC DEVELOPMENT[74]

The earth's surface undergoes a climatic as well as a tectonic development. The earth's climates have changed or at least been displaced and earlier climates have clearly left their mark on the earth's land surface. Their impress upon the landscape of today is sometimes so strong that its character is often more closely related to past climate than to that of the present. But we should not invoke another climate too quickly unless there is also evidence for it in the soil and in plant and animal remains at the same time. Sometimes it is questionable whether particular landforms, gravel terraces, ravines cut into broad valleys or remnant surfaces at high altitude, for example, indicate changes of climate or of level; but changes of climate usually manifest themselves in a different way from those of internal build. With the latter it is largely a question of processes being displaced; in simpler cases they operate at a higher or lower level, in the more complex, old landforms are so disfigured as to be no longer recognisable. Changes of climate produce a different style of landscape. Each climate produces distinctive landforms, and where an earlier climate differed from that of the present, different kinds of landforms exist side by side: the landscape is heterogeneous. Just as a Romanesque cathedral crowned by Gothic features may rise from the foundations of a classical basilica, the landforms of different climates are superimposed upon, and integrated with, one another.

When climate changes, the alterations which take place on the earth's surface are sometimes multifarious and complex. The major processes by which material is transported from one locality to another change: glaciers replace rivers and *vice versa*, rivers carry more or less water, or their flow becomes markedly periodic or they dry up altogether, wind gains or loses in effectiveness. But weathering and denudation also change: frost weathering becomes more or less effective as temperature fluctuates around the zero point, chemical decomposition changes with an increase or decrease in humidity, as it becomes drier insolation increases and with it rock disintegration. Finally, plants and animals also change and in doing so influence, in a manner as yet only partially explained, not only weathering and denudation but even the major processes of transportation. Naturally, a particular climate has to prevail for a sufficiently long time to become effective; short-term episodic changes have little or no effect. It can be assumed theoretically that the longer a climate lasts the more pronounced will be the landscape features peculiar to that climate; Davis, indeed, spoke of a glacial and an arid cycle with this in mind, and Gautier* has recently explained the several kinds of desert in terms of their being of different age. But here again features which differ because they are the result of processes of a different nature or intensity have been wrongly taken to be of different age.

These changes are interrelated in a particular way. In each kind of climate, so long as the internal build remains constant, the processes of transportation, weathering and denudation, as well as the influences plants and animals exert, and therefore surface features and soils, have an individual character and are attuned with one another. Any change in climate will be accompanied by specific changes in the processes and the development of landforms and soil types. It is this fact which requires us to study in our researches the geomorphological significance of each change of climate. We will learn to understand many phenomena by studying processes, and the landforms to which they give rise, in different climates. But I believe it is premature for us to try to deduce a complete geomorphology for each kind of climate and climate changes, as Passarge has done in his *Physiologischen Morphologie*. We do not yet know enough about the processes to do this. Here, too, the need for inductive study is indicated. But this must not be an empirical 'ragbag'; it must be guided by theoretical reflection.

The most recent European climate which differs from that of the present, and one that has influenced very considerably the history of European settlement, was somewhat drier. Though first inferred from plant remains, it was later recognised in a characteristic peat-bog interstratum, the so-called 'limiting horizons' (*Grenzhorizent*). The sand dunes of many parts of Germany may have originated at this time; but apart from this, no other trace of it has as yet been found in the land's surface form.

The loess of Germany is older; according to modern geologists it does not belong to a warmer and drier interglacial period but to the later phases of the Glacial Epoch. They hold that the dust which accumulates to form loess was derived from the deposits of Pleistocene glaciers and ice-sheets or their meltwater, and perhaps from more recent river-flood deposits. The fact that the two are often associated in Germany, and in other loess areas, would support this view, although it is hard to see how the grass-steppe in which the loess dust must have settled can have reached right up to the ice margin.[75] But in precisely those areas of the earth where loess is best developed this association is lacking; its accumulation here is linked much more closely with the fact that these areas are adjacent to deserts, and thus it completely sustains Richthofen's masterly interpretation.* He was wrong only inasmuch as he said that the dust was deposited in deserts without exterior drainage, and later dissected and its salts removed. But dust can be deposited only in an area with a periodic rainfall and a continuous cover of grass; in such an area streams will form and will incise valleys during the rainy season. Obrutschev* has shown that arid deserts do not have loess; on the contrary, it is largely associated with grass-steppes and was deposited in these areas under conditions similar to those which prevail today. It is only when loess occurs in lands which are now wooded that we have to conclude that at the time it was formed a different, drier climate existed.

There are no grounds for asserting that Germany experienced a desert climate in the recent geological past. The peculiar rock forms of the Quadersandstein and in part the Bunter Sandstone too, on the basis of which this has been claimed, are not the result of desert conditions; they are due to the porosity of the rock. Such an example of convergent development has deceived many research workers.

The Glacial Epoch is the most familiar and best studied manifestation of climatic development. Massive ice-sheets repeatedly covered the greater part of the land far into the temperate zone, and even in the equatorial zone higher mountains carried snowfields, from which large or small glaciers descended into the valleys; the climate was colder everywhere, and the plant and animal life adapted to the greater coldness. But the glacial periods are to be seen as episodes, even if phenomenal episodes, in the development of the earth's surface. During interglacial periods as well as earlier, particularly at the close of the Tertiary and the beginning of Quaternary times, the climate was as warm, perhaps warmer than it is today. At such times, rivers and rainfall would be the dominant forces shaping the surface, as they are today. The glaciers filled valleys which already existed, or mantled the hills and plains shaped by fluvial action. To understand the configuration of the surface today we must seek to appreciate not only the topography of glacial times but that of pre-glacial times as well. The question now actively debated is whether the bottoms of Alpine valleys had come close to the profile of equilibrium when glaciation began, or whether river erosion was in full spate; in other words, are certain landforms to be attributed to glaciation or to the preglacial work of rivers? The issue will scarcely be settled by studying the Alps alone; a comprehensive comparison with unglaciated high mountains in warmer climates is called for. When there have been several glaciations, each separated by warmer interglacial periods, we also have to discern the effect each glaciation had in shaping the surface. Indeed multiple glaciation was first recognised as a result of distinguishing between multiple gravel terraces. But very little work has as yet been done along these lines in the interior of glaciated regions where excavation predominates; Lucerna's assertions* are still forcefully contested by other glaciologists.

Although opinions still differ greatly as to exactly how particular landforms are developed by glacial erosion and glacial and fluvioglacial deposition, their great importance is now generally recognised.

On the other hand, not until recently has any great attention been paid to the influence of the colder climates which prevailed during glaciations in areas far removed from an ice-cover. On peaks rising above the firn and glaciers, and in an extensive zone bordering glaciated areas, in the periglacial regions as they have been called, weathering and mass-movement must have been different from that of today and more akin to that of the polar zone tundra: frost weathering and solifluction must have played a greater role, and wind may also have been more active; the majority of geologists consider the material forming the loess of Germany to have been derived from bare moraines and fluvio-glacial gravels. But since the onset of such colder conditions involves a change in the intensity of weathering and deundation rather than in their nature, it is difficult to distinguish between periglacial sculpturing of the surface and that of today. As only too easily happens with new ideas, we have probably gone too far in claiming that periglacial processes have been at work, and ascribed to the Glacial Epoch such features as block streams and boulder fields when they can still be formed today.

In the deserts and semi-deserts covering such large areas of sub-tropical latitudes Frass, Hull, Gilbert* and other research workers have found evidence of a past moister climate; they termed it a pluvial climate and considered it contemporaneous with the glaciations of higher latitudes. It is true that Walther, the distinguished student of desert geomorphology, disagrees; but in this case, misled by his loving absorption in the study of present-day desert processes, he has most certainly misapplied a research principle, in itself correct, of invoking different conditions in the past only when present-day forces have actually shown themselves inadequate. That the climate of at least the northern margins of sub-tropical deserts has changed is an incontrovertible fact, as Blanckenhorn* stresses. It is impossible for wind to have deflated the elongated hollows, with their recurring contrast of undercut and slip-off slope and in fact all the characteristics of valleys; the cloudbursts of today could scarcely have formed the valley systems of such length and ramification as are to be found in many parts of the Sahara and Arabia. What significance gravel terraces lying at the mouths of large wadis on the edge of the Nile valley have is debatable. Blanckenhorn saw them as features formed in a moister climate; Passarge,* on the other hand, thinks that they must rather be seen as features of a drier climate, their erosion as characteristic of a moister climate. Like the terraces on the margin of the Upper Rhine trench, they may be the result of progressive subsidence, not of climate change at all; but this is certainly still open to question. During periods of moister climate the water level of lakes in regions of interior drainage will rise, sufficiently perhaps even to establish an outflow. Since Gilbert and Russell* discovered the elevated strand terraces and former outlets of the large lakes of the North American Cordillera, similar phenomena have also been found in other regions of interior drainage.

As yet we have few observations on changes in the climate of the tropics during Quaternary times, apart from those on the lowering of the snowline and glaciers extending further down-valley. Probably, temperature was in general somewhat lower. The possibility that the seasonally-moist steppe and savanna climate, or conversely the permanently-humid forest climate, became more extensive than they are today must also be borne in mind. Lang* believes that laterite is formed in periodically-moist climates, not in those permanently moist; he maintains that there has been a change of climate in areas where laterite is now found in a climate permanently moist.

By the end of Pliocene times, the climate appears to have become similar to that of the present. But fossil flora and the present distribution of plants force us to conclude that the climate of higher latitudes was appreciably warmer during the Miocene. Heer* maintains that at this time South Greenland had a climate similar to that of Carolina today. The snowline of mountainous regions must have been higher; at a height where there is now firn and glaciers, valleys could be incised by running water; valleys later filled by glaciers probably date from a period following upon the major processes of mountain building. On the other hand, limited areas of middle latitudes appear to have become drier than they now are; according to Penck's investigations,* central Spain was a region of interior drainage comparable with that of the Schott region of the Atlas lands and interior Asia Minor today. This is indicated not only by the gypsum and saline deposits of the plateaux, but also by the peculiar rock platforms at the foot of the Sierras of Castile, analogous only with those of the seasonally-moist tropics of today. Central Spain seems to have become a region of peripheral drainage only recently. Penck believes that a moister tropical climate encroached upon the present equatorward boundary of the arid zone. The red laterite found by Walther* in the areas bordering the Egyptian Sudan and the Sahara may belong to this period.

The climates of still earlier times, those pre-dating about the middle of the Tertiary period, are generally of little relevance to our understanding of the present; the surface configuration of these times has been largely obliterated by subsequent large-scale dislocations. Only where large land masses have since remained stable have the configuration of Lower Tertiary and Mesozoic times been preserved. Passarge* attempted to explain the pediment and inselberg landscape of tropical and sub-tropical Africa in terms of a desert climate prevailing throughout Mesozoic time; but inselberg landscapes occur in all parts of the earth where there are seasonally-moist climates, and are more easily understood as features of such a climate than of arid areas.

An important fact stands out when we take an overall view of the changes of climate in more recent times. There is as yet no reason to suppose that climates quite different from any which now exist have existed in the past; on the contrary, it seems to be a matter of gradual transitions of climate or—what may in part amount to the same thing—of climate regions being displaced; climates were of a more polar or more equatorial, a more oceanic or more continental character than they are today. To trace these shifts and changes is an important task for geology and palaeogeography. But in doing so these disciplines should not limit themselves to studying individual climatic factors, as they usually have done hitherto; to learn its nature and causes, climate must be treated as a whole. Koppen and Wegener* were the first to try to do this in any comprehensive way; but perhaps their effort rested too heavily upon a particular theory.

We do not yet know with any certainty what causes either the processes on which internal build depends or changes in climate; for the present, therefore, we cannot deduce what changes should follow from such causes. Fossil flora and fauna and plant and animal geography afford us some help in establishing past climates; the results of geomorphological or pedological studies carry more weight when they are confirmed in this way. Thus Richthofen's theory on the origin of loess was more readily accepted after Nehring* had come to the same conclusion from a study of the animal remains which it contains. Heer's* study of the fossil flora of Greenland advanced our knowledge of the warmer climates of the Tertiary. But the task seems to devolve mainly upon landform and soil studies. Although they were late in starting, such studies have already produced important results. But we have not always been careful enough, and as a result reached false conclusions. Just because the surface features we find in an area also occur in, or are even in large measure limited to, desert or polar climates, we are not entitled to immediately conclude that the area once had a desert or polar climate. We must first make an exhaustive examination to see whether the same features could not be formed under present-day conditions. The ways in which features might now be formed must always be examined first; only when these prove inadequate is it permissible to invoke different conditions in the past. In particular it is comparatively rare for the minor landforms of an entirely different past to have survived weathering; such past conditions are usually reflected only in the larger landforms. Nor should we one-sidedly overemphasise either tectonic or climatic change; both possibilities must always be kept in mind. For although one can usually distinguish the effects of tectonic change from those of changes in climate—the former affecting the level, or storey, at which erosion and degradation take place, the latter their style—the difference between the two is by no means radical. In explaining many major and minor landforms one can still be doubtful whether they are the outcome of changes of level or of climate. If we work deductively both explanations seem possible; only inductive study will be decisive.

CHAPTER X

The Geomorphological Interrelationship of Landscapes

PROCESSES BY WHICH MATERIAL IS MOVED

Two types of process modify the earth's land surface: weathering, be it the mechanical disintegration of rocks or their chemical decomposition, and the processes of movement and relocation, of removal or degradation, of transport and deposition. In reality these two types naturally go together: relocation is always associated with weathering, if only because of the effect of gravity, and conversely relocation is always combined with changes in the material, for example rock comminution and rounding. But they are different in their nature, depend on different causes, and work in different ways to shape the earth's surface. They differ in one essential way: weathering processes are determined by conditions in the particular area where it is taking place; but it is of the nature of the processes whereby material is relocated that one area is related to another. All degradation must necessarily result in deposition elsewhere, and this in its turn affects degradation. Thus areas interact or are correlated with one another.

Since the processes which relocate material are ubiquitous, no part of the earth's surface is quite autonomous and entirely independent of other places; on the contrary, each part must be seen as a part of a greater complex or system, or region of the earth, as far as surface modification is concerned. True, the extent and scale of relocation differ from place to place. This can be so within quite a small area, for example, on a mountain side, where gravity, rain and soil moisture or avalanches bring down material loosened by weathering and pile it up at the foot. Relocation in this case is the result of height difference; neither climatic nor any kind of regional contrasts are involved to any appreciable degree. Naturally, débris which differs in its material and its surface form from its bedrock and its weathering mantle is more prone to movement; but such movement of material is still fairly closely linked with weathering, and since it depends on the same conditions of climate and rock, we can consider the geographical distribution of weathering and of denudation together as a unity. The relocation of material plays a special role when it involves material being brought from one area to another, and in this regard different altitudinal zones of a mountain region are to be seen as different areas. In this way one area becomes dependent on another, so that it can no longer be understood in terms of conditions prevailing within it alone, but only from its relationships with other areas. A. Penck[76] accordingly distinguishes autochthonous and allochthonous phenomena; Passarge, who formerly spoke of consonant and dissonant phenomena, recently proposed[77] the expression native forms (*Heimatformen*) and foreign forms (*Fremdlingsformen*), but I prefer local (*orsständige*) or indigenous (*eingeborene*) and alien (*ortsfremde*) forms; for the word 'native' (*Heimat*) has a special meaning.

We know of three sub-aerial processes of movement and relocation on a major regional scale, namely glaciers, rivers and wind, coastal processes again excepted as they were in preceding chapters. Thus we can distinguish between movement and relocation by glaciers, rivers and the wind, between glacial, fluvial and aeolian movement and relocation.

Where these three kinds of movement and relocation occur and predominate depends on climate, for they originate in different climate regions. But their dominance is not conterminous with these climate regions; it generally extends beyond one region into adjoining regions.

Three principal types of climate are involved.[78] In firn or nival climates the summer heat is insufficient to melt winter snowfall, which therefore forms as a permanent cover. In moist (or humid) climates the heat is sufficient to melt the winter snow, but evaporation and percolation together are not able to dispose of all the solid or liquid precipitation. In dry (or arid) climates, evaporation not only exceeds and disposes of the annual average precipitation; it is greater than precipitation throughout the year.

These climate types are not genetically different. They are quantitative transitions and separate in rather a fortuitous way, climates of one and the same character. But they are of the greatest significance for the water régime and surface configuration. Areas of nival climate are the source regions of glaciers; but glaciers extend, thanks to their movement, into warm-moist climates. Areas of permanently or periodically-moist climates are the source regions of rivers; but these can flow into dry zones. Only ephemeral streams can originate in permanently-dry desert climates; rivers which flow permanently or periodically must necessarily have come from outside such areas. It is in such desert areas that besides occasional cloudbursts the wind becomes an active agent in surface sculpturing; it lifts sand and dust from the bare unvegetated ground and carries it away. The desert is the source of material relocated by aeolian action; but only sand is deposited there, dust being blown away to settle only in adjoining areas. Thus glacial, fluvial and pluvio-aeolian shaping of the surface in no way coincide with nival climate, moist climate and dry climate. In its source area a glacier is subjected to a nival climate, to a moist or dry climate where it wastes away. A river basin need not be wholly within a moist climate, but may have a moist source area and a dry wasting area. In the transportation of material by the wind, deserts act as source areas, periodically- or permanently-moist lands as receiving areas.

With each of these three types of movement of material, the processes of degradation, transportation and deposition take place in different ways. It is for this reason that it is of first importance that the processes should be distinguished by geomorphologists. But while it is true that there is no area in which degradation and deposition do not occur together, in some areas degradation predominates, in others deposition; areas in which the two are in equilibrium are rather exceptional.

Firn and glaciers consist of three basic forms according to their role in the landscape. The small areas of firn which develop at about the altitude of the snowline in the upper end of valleys and with which only small glaciers are associated are limited to crests and summits; they only modify the fluvial landscape. The larger firn fields lying on plateau surfaces, or in the funnels and hollows of source areas, form plateau or valley glaciers, which may remain within the mountains or descend out onto the foreland. Ice sheets, which differ from plateau and valley glaciers not only in size but also in the fact that their vertical component of movement is much less than the horizontal, blanket whole lands. When compared with rivers all glaciers are slow moving and of great thickness.

The processes of glacial degradation and deposition are still by no means explained; naturally, they are largely buried beneath glaciers, and they can be directly observed only on the land once glaciated but now free of ice. As with rivers—but we do not yet know under what conditions—glacial degradation and deposition do not seem to be sharply separated either in time or space; but on the whole degradation predominates in the higher parts of glaciers and beneath them, deposition in their lower parts and beyond them. Perhaps it will in time be possible to develop a schema for glaciers corresponding to the profile of equilibrium of rivers, and from this to deduce the processes of degradation and deposition. Till now we have scarcely begun to do this. Indeed, nowadays many research workers again dispute the efficacy of glacial erosion altogether and attribute to glaciers no more than the ability to plane off and smooth valley floors. The much greater thickness of glaciers means that their beds are much larger than those of rivers—glaciers can only be compared with the beds of rivers and not with their valleys, as Penck strongly emphasises. The weathering and denudation by which the sides of river valleys are shaped play a much smaller role in glacial valleys, limited as they are to the dividing ridges. The landforms of glacial degradation vary with the kind of glacier involved; corries develop in the firn region of small glaciers, valley glaciers bring about the glacial modification of valleys, plateau glaciers and ice-sheets affect both upland plains and lowlands.

Moraines, whether ice-surface, ground, or terminal, are the hallmarks of glacial depositions. But the oser (*Aser*), drumlins, fluvioglacial gravels, etc., formed beneath or at the margin of glaciers by melt-water deposition follow closely behind. The deposits of small corrie glaciers extend only a short distance down mountain sides; those of valley glaciers lie in the valley or in front of the margin of mountains; those of large ice sheets mantle entire plains and not necessarily ones lower than the area whence the material has come. They form morainic girdles (*Gürtel*), often at the same height above sea level. When glaciers issue into the sea, waves seize and remove the glacial débris. On dry land broad gravel plains formed of meltwater deposits lie in front of the moraines. But one must be careful not to rashly ascribe to all such gravel plains a fluvio-glacial origin;

whether, for example, the gravel terraces of the Upper Rhine plain do in fact correspond to the several glaciations still needs more detailed examination.

The configuration and make-up of the earth's surface is not only a function of the present-day snowline and glaciers, but even more of the snowline and the more extensive glaciers of the Glacial Epoch; for the features and deposits then formed, or at least those formed during the Last Glaciation, have since been only slightly destroyed; for the most part they survive today to give their distinctive impress to the landscape. The surface features and soils of the whole of North Europe as well as the northern part of North America are the result of degradation and deposition by the Pleistocene ice sheets. In the mountains of lower latitudes, glacial landforms occur at a much lower altitude than do the glaciers of today. Glacial landforms and soil formed in the Pleistocene are found in areas where landforms and soils are now of fluvial origin. Bare rock and morainic landscapes developed during the Last Glaciation are still more or less intact. By contrast, those of earlier glaciations are markedly degraded, but can be seen to be of glacial origin from the material of which they are built if not from their form.

FLUVIAL TRANSFERENCE OF MATERIAL

At the lower end or beyond the margin of Pleistocene glaciers we pass from the glacial to the fluvial landscape. In lower latitudes and in dry climates this transition takes place at great altitudes above sea level, in higher latitudes only at lower altitudes. When the land does not rise as high as the snowline the landscape is entirely fluvial. Only in dry regions where evaporation exceeds rainfall at all seasons does the fluvial landscape again disappear.

The different discharge (*Wasserführung*) of rivers in different climates naturally influences the working capacity of rivers, and the processes of degradation and deposition. But this influence is probably less than one might at first think; for it seems to depend on the discharge at the time of high water rather than on the total annual discharge. Only when the river flows for too short a time is it unable to keep its course open; we are then in streamless dry landscapes.

Just as the areas where snow accumulates, because it never entirely melts during the summer, form the source regions of glaciers, so rivers are formed in areas where rainfall exceeds evaporation, at least during one season. As long as a river flows in a moist land, rainfall and tributaries increase its volume. Where the moist climate reaches to the sea so will the rivers, to fall within Richthofen's term 'peripheral'. Even when depressions interrupt their course, rivers soon fill these up and flow out again. Only in karst mountains do rivers sink into the rock, and though they reappear elsewhere they have in this case an indirect outflow; karst mountains occupy a position intermediate between areas with drainage to the sea and those without.

If, on the other hand, a river enters a dry region in which evaporation exceeds rainfall at all seasons, its water volume decreases downstream instead of increasing. Rivers waste away in dry regions. Small rivers will rapidly dry up completely and larger ones are diminished; how long they can maintain themselves against drought and whether they still reach the sea, like the rivers of the West coast of Peru, and the Colorado, and the Nile, or dry up before they do so, like the Amu Darya, Syr Darya and many others, depends upon their volume and on the extent of the dry region they have to cross. After Richthofen, we call rivers which come to an end before reaching the sea, interior rivers. If all the rivers of a region dry up we can call the whole of the area an area of interior drainage (*Zentralgebiete*), in contrast to a peripheral area. If on the other hand, only the smaller rivers dry up, the larger flowing through it, it is better to speak of areas of semi-interior drainage (*halbzentralen Gebieten*). Thus areas of interior drainage are by no means always dry regions throughout; they are often made up of a moist source region and a dry one in which rivers waste away. The part of central Russia drained by the Volga and its tributaries belongs hydrographically to an area of interior drainage, as it is often carelessly said to be, but merely an area without outflow as it is often carelessly said to be, but merely an area without outflow to the sea; it certainly is drained, and it is relatively unimportant what happens to the outflow.

In the case of peripheral rivers in moist regions the profile of equilibrium is concave and begins at the coastline. When high land abuts on the coast, even the smallest river can incise throughout the whole of its course and form a valley landscape. If, on the contrary, lowlands back the coastline, and these lowlands are lower than the profile of equilibrium, any débris brought down by rivers from higher ground farther inland will be deposited on the lowland and in the sea.

The Geomorphological Interrelationships of Landscapes 121

Rivers also aggrade in inland depressions. Deposition can in this way take place even at a considerable height above sea level, because the profile of equilibrium, in keeping with the greater distance from the sea, is higher, and stretches in which the river aggrades alternate with those in which valleys are incised. But when the whole river basin is lowland and lies lower than the profile of equilibrium throughout, neither valley development nor aggradation takes place, for neither is possible if the river does not erode upstream and carry débris down with it. The rivers of tablelands also must remain static with erosion at a standstill, until headward erosion reaches them.

In the case of peripheral rivers with their lower courses in dry regions, vertical erosion ceases even at a considerable height above sea level because of the steeper profile of equilibrium; vertical erosion is absent at an elevation where, in a moist climate, valleys would still be formed. But where the land has an appreciable relief, rivers from moist regions entering even the driest desert excavate valleys with marked canyon characteristics, as they do in the Cordilleras of North and South America.

Even interior (*Zentrale*) rivers can erode in their upper courses; but towards their lower courses their power to transport progressively declines and finally ceases. They have to deposit all their débris. Deposition is greatest where they emerge from mountains on to a plain, when a sudden fall of gradient coincides with decreased discharge. Here large talus cones develop, the gradient of which decreases away from the mountains, and since the gradient decreases outwards so in the same direction they are made up of finer and finer material, loam giving way to clay. Compared with rivers discharging into the sea they have been termed 'dry deltas'. Here the rivers generally anastomose into numerous channels, which increases evaporation still more, and continually alter their course. Where they discharge into enclosed basins, either of tectonic origin or deflated by wind, they form lakes, the position of which frequently changes; as existing hollows are rapidly infilled, new hollows develop nearby. Unlike all kinds of deposition in moist regions, brought about by decrease of gradient alone, aggradation in interior dry regions is caused by evaporation; even the salts carried in solution are deposited. Since the ground is bare as a result of the dry climate, the wind winnows out the finer particles to form dunes elsewhere. Such interior regions of deposition may be formed at very varied heights; they are found as depressions below sea level as often as at very great altitudes. And indeed, without any change of climate, their drainage can be captured from beyond the limits of the depression and absorbed into the peripheral drainage system.

Just as with an increase in temperature and a decrease in snowfall, glaciers and glacial landscapes give place to rivers and fluvial landscapes, so with greater aridity they give way to streamless landscapes, where degradation and deposition coincide with occasional cloudbursts and wind. These are deserts and semi-deserts, for only in such areas do cloudbursts and wind play this role. Not all deserts and semi-deserts belong in the fullest sense of the word to this group, only those which are streamless, or those in which rivers are few and far between; we can call them independent deserts in contrast with dependent ones which, by virtue of their rivers, are related to moist areas. The extensive areas of low desert tablelands and remnant desert landscapes are generally independent; the deserts of mountain ranges and highly eroded block landscapes, the higher parts of which receive a more plentiful precipitation and give rise to rivers, are dependent, and therefore areas of dissection or aggradation. The forces peculiar to deserts by which material is relocated are effective in every desert, and begin to work whenever running water lays down the particular tools with which it carves the landscape.

Two forces which relocate material are involved in the formation of desert landscapes: cloudbursts and wind. The major importance of wind for the desert landscape has been particularly stressed by Walther. But in exaggerating the role of wind it often sounds as if he denies running water much of its effectiveness, especially in the formation of wadis. Now he has declared that in this respect he has been misunderstood. The work of running water is so evident in the wadis characteristic of many desert areas, especially in the constantly recurring contrast of undercut and slip-off slope, that a geomorphologist just cannot question it. And as we have seen, these wadis can hardly be ascribed to the occasional downpours of today. They probably indicate a moister climate in the past; episodic or ephemeral streams waste away or sink quickly into the ground, and constantly alter their course.

Wind degradation and cloudbursts together produce rock deserts out of bed-rock; according to internal build these will be mountain deserts or tabular hamada. Deposition by cloudbursts produces gravel and loam deserts, and by the wind, sand deserts. On the other hand, dust blown away from deserts is not redeposited until it reaches neighbouring steppes, in other words a fluvial landscape. Loess landscapes are not areas once dry into which rivers later penetrated, as Richthofen at first thought; they are the result of dust deposition and river erosion interacting in both time and space.

CHAPTER XI

Coasts

Coasts are so closely related to the land surface that we cannot afford to disregard them, especially since in tracing the development of their study we can also trace the development of geomorphology as a whole.

Their geographical study was also purely descriptive for a long time. Since they are at sea level their horizontal features were given special importance and they were therefore usually studied from three different viewpoints: their ground plan, their elevation and the rocks of which they were comprised.

Such an analytical approach certainly affords a great deal of valuable knowledge; but it remains essentially unsatisfactory. Coasts, like landforms as a whole, are individual features. Each stretch of coastline can and must be considered as an entity in which its characteristics are combined; coastlines with similar characteristics must be grouped together under a single type name. Seafarers, and those living on the coast, have long since developed such a unified approach and expressed it in language later adopted by geomorphology. The best known example of this is the word 'fjord'. Of Norwegian origin this word has been applied to similar coasts in other countries, and thus become a general type name. It was inept that many teachers later extended it to include all rocky inlets, since it then lost its characteristic meaning and left us without an appropriate term for such glacially-modified inlets; it is more expedient to leave it with its particular meaning. The bays of the north coast of Spain are called *rias*; when Richthofen became acquainted with similar bays on the southern coast of China he transferred the name *ria* to them, and to all bays of similar shape. In different parts of the German Baltic Sea coast the bays are called *förden, bodden* and **haffe**, and these names too have been adopted to some extent. In this way a large series of bay types has gradually found its way into colloquial and scientific language. Since particular forms usually occur in a group, all the bays of a coastline being of a similar type, the type name can be transferred to the whole coastline. We can speak of *fjord* coasts, *bodden* coasts, *ria* coasts and so on. Richthofen was able to draw up a table of coastline types.[79] And at almost the same time F. G. Hahn could base his work on transport geography[80] on a series of coast forms; though named after the country in which they most characteristically occur, they are roughly equivalent to Richthofen's. These coast types are at root descriptive; but since they take account of all the characteristics of coastlines they already represent a natural rather than an artificial classification.

The study of coastal processes progressed for a long time independently to that of coastal landforms. Coastal landforms were the stuff of geography. Coastal processes were the concern of geologists, who, especially in insular England, devoted themselves energetically to the study of coastal destruction. Hydraulic engineers, from practical considerations, also had to be concerned with the processes, and advanced our understanding of many phenomena. This knowledge of forces and processes was an indispensable prerequisite for a genetical interpretation.

THE STUDY OF COASTLINE PROCESSES

Observations and studies of secular uplift and subsidence or shoreline displacement were also of great importance. These had been noted as early as the eighteenth century along the Scandinavian coast; later they were recognised on other coasts also. But they were often studied uncritically, and maps of their distribution brought confusion rather than gain. Only after Suess's conclusive critique* had cleared the decks and established more clearly the facts of uplift and subsidence, could they be introduced without misgivings into the study of coastal landforms.

The explanation of coastal forms, like that of landforms, began with vague speculation. One of the oldest scientific studies of coastal forms is that on fjords by the American naturalist Dana,[81] a study which, early though it was, was correct in its essentials. Peschel* introduced the genetical approach to the study of coasts into the German literature with his paper on fjords. Compared with Dana's paper, of which he was ignorant, Peschel's was factually retrograde, since he invoked fissures to a much greater extent. But though he did not recognise that fjords are related to the former distribution of glaciers he saw that they are limited climatically. His study of deltas followed a similar path (the treatment by Réclus* in his work *La Terre* was of approximately the same date) and led to Credner's more fundamental study* though one still based entirely on the comparative study of maps and literature. The fine study of the coast of Brittany by the Basle anatomist Rütimeyer* did in fact arrive at a false factual conclusion. But like his earlier studies of valley and lake formation it had the merit of bringing about acceptance of direct observation, and this soon became the mainstream of geomorphological method. Even Richthofen could, for the most part, give his coast types a genetical connotation. And it was not too early for him to make a start on a truly genetical classification, that is to say, one based on origins. He distinguished two major classes of coasts: those where the sea has penetrated into the land, and those along which there has been alluvial accumulation. We have progressed farther in this direction. I might cite Philippson's study of coast form types, especially those of alluvial accumulation,[82] at the conclusion of which he sets out a genetical table distinguishing in particular potamic (*potamogenen*) and thalassogenous (*thalassogenen*) alluvial coasts. During a journey in the autumn of 1898, the genetical relationships of the Norman and Breton coasts became clear to me, completely different as they are in their horizontal dissection, but so similar and genetically related in their elevation. Both have been formed by wave attack following upon subsidence of the land; but while in the weak chalk of Normandy the attack progressed rapidly and fairly uniformly, and could hardly result in the formation of bays, in the hard material of Brittany it has worked more slowly, resulting in the submerging of the land; waves gnaw tumultuously at the promontory cliffs, but adjoining river valleys have been converted into deeply-embayed inlets. On this basis one can generally distinguish, even if somewhat imprecisely, between smooth rocky coasts where aggression exceeds ingression and embayed coastlines where wave attack is no less strong but is exceeded by the penetration of the sea into the land. Certainly, according to their shape in the past, coasts develop different forms. For example, the fjord coast is the type-form of a glacially-modified mountainous or plateau coastline, the ria coast is modified not by glaciers but by fluvial action alone, where valleys transverse to the grain of the country predominate. In this way it will gradually be possible to deduce all coastal forms from the conditions in which they occur, from the structure and shape of the land, the presence of uplift or subsidence, tidal range, wave attack, coastal currents and so on. But deductive arguments have scientific value only when they are tested step by step against the facts.

Naturally, among the factors that determine a coast's form is its age, or put more precisely, its stage of development. Philippson took this into account when he distinguished between incomplete and complete potamic alluvial coasts. Davis and his followers also stressed the age of coasts. The first attempt in this direction was made by Gulliver* in 1899; it does not seem to have been accepted by his own school and seems untenable to me. Later Davis himself, Braun and others framed a deductive schema for coasts based on age; recently D. W. Johnson[83] also put age at the centre of his otherwise very fundamental treatment. To take only one example, while the different development of the Norman and Breton coasts seems, as we have said, to be the result of different rock resistance, Johnson interprets smoothly-cliffed coasts as an advanced stage of development of embayed rocky coasts. That makes sense, however, only if it can be established that the Norman coast was also once an embayed rocky coast, and has lost this form possibly with the progress of destruction. But a plateau coast composed of rock as weak as chalk, and at the same time as little dissected as that of Normandy, can never have been an embayed coastline. The two types of coast are not successive forms; they develop alongside one another simultaneously, because the rock or other formative conditions differ. Thus Davis's development-stage approach again misfires when it is applied to coastlines.

At one time, regional geographies devoted a special section to coasts on the trivial grounds that travellers arrived at the coasts of a foreign land first; they did not relate coasts to the overall nature of the land. The genetical approach will bring coastlines into a close relationship with interior landforms. The overall form of coasts must, as Richthofen put it, be linked with the physique (*Plastik*) of the continents, that is to say with the major features of internal build. To subdivide coastlines, at least to distinguish coastlines of aggression from those of ingression, the result of submergence of the land is meaningful only when it is related to the subdivision of the lands. The kind of surface modification, fluvial, glacial or pluvio-aeolian, to which the land is subjected is, like its structure, reflected in the form of its coastlines.

CHAPTER XII
Theories on the Origin of the Land Surface

In all sciences which, not content with mere description or incidental explanation, aim at a comprehensive exposition and causal association of a whole complex of facts, hypotheses and, with their scientific development, theories, play a major role. Occasionally, the causal relationship between phenomena can be traced directly to a single process; but this can usually be arrived at only after much thought. We begin with a hypothetically valid general proposition on relationships, and we test it against the facts. If the supposition is substantiated it becomes a theory. As examples I need only cite physical or chemical theories, the theory of evolution or the theory of egoism, or the more recent theory of diminishing returns in economics, or the theory of historical materialism, each of which seeks to explain a whole field of phenomena with greater or lesser success. In the same way, as soon as geomorphology becomes more than morphography, it must not content itself with description; it must explain the landforms of the earth's surface, and understand their distribution and relationships; it must start with general theoretical propositions as to their origin and cannot succeed without the use of theories.

Insofar as the ordinary person considers landforms at all, which he rarely does, he conjures up violent events, great floods, volcanic eruptions and earthquakes, particularly when he is confronted by impressive landforms. Budding science scans wider horizons and sees the overall surface of the earth as a problem in itself, and puts forward comprehensive explanatory theories; but the content of its approach is still essentially the same. The Neptunist theory used major floods, whereas the Plutonist theorised reaction in the earth's interior which manifested not only in volcanoes, earthquakes and mountain uplift, but also in the formation of open fissures, and valleys. The two theories agree in that they think of these as mighty events, as catastrophes or revolutions, annihilating the whole set-up and thus clearing the way for a new one. Such a stage of scientific development, one which employs imaginary, even non-existent forces and processes, is in fact only a preliminary step to science.

The true science of geomorphology did not begin until revolution was replaced by gradual development or evolution, consisting not of mighty, unknown natural events but of the summation of processes daily operating before our eyes. Precursors of this actualist approach are the two Frenchmen, Guettard and Demarest, the Scots, Hutton and Playfair, and the Thuringians, Heim and von Hoff.* But Charles Lyell first brought it into general currency, and for this reason he has considerable importance for geomorphology, even though he exaggerated the effect of the sea, and underestimated valley formation to an inhibiting degree. Truly scientific approaches and theories are possible only on the basis of actualist and evolutionary theory.

THE SCIENTIFIC DEVELOPMENT OF GEOMORPHOLOGY

We can distinguish three periods in the scientific development of geomorphology.

During the earliest period the approach to the earth's lands continued to be Plutonist: the earth's surface is seen as basically the work of forces in its interior. Modification of the surface by water or indeed any sub-aerial forces is of secondary importance. The upwelling of hot liquid magma, reaching the surface partly as volcanic eruptions and partly by pushing against the rigid crust of the earth, and thus lifting it and thrusting it sideways, was for a long time the only way the earth's interior was thought to have its effect. As late as the 1860's, theories of mountain building were dominated by this conception.

It was seen that the sea acts upon coastlines, so the effects of the sea extended far inland as well; only here and there was an attempt made to assess the effects of flowing water and thus to explain the origin of valleys. Most research workers saw these as fissures caused by uplift. Only small-scale weathering and degradation were recognised as effective inland forces; they produced soil and minor landforms.

In the 1860's and 1870's, and in English geology somewhat earlier, a change occurred at roughly the same time both in the approach to tectonics and in the appreciation of sub-aerial forces. More detailed field work and the gradual accumulation of observations can be seen as the common cause of these changes. From then hot liquid magma was seen as an active agent only in the case of volcanic eruptions. Mountain building tended to be attributed to more general processes in the earth's crust, which affected even crystalline rocks. Folded and block mountains were distinguished, and thus the way opened for the scientific analysis of the internal build of the earth's land surface. Fissures were posited only where they could be observed; they had played out their role as a way to explain valleys. On the other hand observation of rivers, as well as of the marked effect of running water on incoherent material, led to greater efficacy being attributed to rivers themselves; observations of elevated valley terraces explained as old valley floors, provided evidence for the erosional nature of valleys. But only these and minor landforms were seen as the work of surface modification; the surface of tablelands and plateaux was thought to be the original surface, and even in the case of mountains, only small-scale dissection, not general degradation and lowering, was thought to be significant.

After the way to an appreciation of the processes of surface modification had been opened up, it was impossible for knowledge of these processes not to be progressively advanced, impossible for the denudative processes and advanced degradation and modification not to be recognised as well. The drawing of geological profiles showed ridges where the synclinal structure would once have led one to expect a valley, and valleys where the anticlinal structure would have suggested a ridge. Certainly, it was still common for advanced degradation, especially by sub-aerial forces, to be accepted only with reluctance; but increasing knowledge supported such a view more and more, and turned it into a certain scientific fact.

Another piece of scientific knowledge went hand in hand with this. The deposition of recent unconsolidated formations on top of bedrock, in highlands as well as lowlands, had long been noted, but were usually held to be marine deposits. Now these deposits are recognised as those of rivers, wind, or glaciers and ice sheets, in other words as sub-aerial features. Thus powerful aggradation on the land contrasted with powerful degradation; the modification of the land surface by its own forces assumed immense proportions.

New ideas in tectonics, such as the nappe theory of the Alps or the recognitions of widespread uplift, have also provided geomorphology with new approaches and problems, of which we have not yet formed a clear enough conception.

In this way the foundations of geomorphological theory developed since the 1880's, to become the basis of our approach today. Neither internal build nor surface modification is by itself the determinative or even the predominating factor for the earth's surface form; on the contrary, the two always work together. In areas where degradation predominates, the actual surface is the difference between, on the one hand, the tectonic surface as it would have been were it the result of internal build alone, and on the other, of degradation; in areas where deposition predominates surface processes are additive, laying a veneer over the internal build and the tectonic surface. But when we think in this way we must remember that internal build and the tectonic surface are not the result of internal or endogenic forces alone; for the rocks formed under submarine or sub-aerial conditions by surface processes before the decisive folding or faulting provide most of the material with which endogenic forces work. In studying surface modification, geomorphology accepts only the processes operating during and after mountain building. Yet this, as is being increasingly realised, is not usually a unified event; it has often been divided into several phases, each of which took place gradually, not suddenly. Nevertheless mountain building is hardly an uninterrupted process varying only in intensity, as a more recent hypothesis would have it.

So long as geomorphological research was carried out mainly in European countries, and in the temperate zone where differences of climate were of little moment, it could culminate in the knowledge that surface configuration was dependent on internal build; it was for this reason that geomorphology only gradually freed itself from the purely tectonic approach, and allowed for the effect of surface modification as well as for internal build. To start with, the influence rock type and rock disposition have on the surface was recognised only in isolated instances; only gradually did the influence of internal build as a whole come to be appreciated. From the geographical viewpoint the arrangement and distribution of surface features seemed to be completely determined by the arrangement and distribution of endogenic forces, the laws of which, though as yet little known, are in any event based upon the earth's internal make-up, and independent of climate.

But when research extended beyond the boundaries of Europe, and even within Europe, surface forms were seen to display differences other than those related to internal build, differences related to past or present climate. The distinctive landforms of the High Alps and the deposits at the foot of the Alps, as well as the Quaternary deposits of northern Europe extending far into North Germany, turned out to be the product of a period of colder climate, and a more extensive permanent cover of snow and ice during the so-called Ice Age.* And such features were also found in North America, and in all high mountains of the earth, even those of the tropics.

For this reason glacial landforms, though of the greatest variety among themselves, contrast sharply with most fluvial landforms, and like them have a particular geographical distribution. In arid climates also, landforms are different from those we are accustomed to in Europe or in the Eastern United States. To begin with, the canyons of the American West were recognised as the features of an arid climate; then followed Richthofen's loess theory and his theory of aggradation in regions of interior drainage, then Walther's studies in the Egyptian desert, which were especially important for our understanding of minor landforms and our appreciation of degradation by wind. In this way the surface configuration of deserts and steppes also came to be seen as something distinctive, with yet another geographical distribution. Later, the characteristic differences of landforms within the tropics, that is to say, the differences between permanently moist tropical lands and those with a marked dry season, were recognised. And in areas like the Mediterranean, with its dry summer, Philippson* recognised many distinctive features. Thus there is a climatic distribution of landforms as well as a tectonic one. And it is clear that because of their differences climates are effective in different ways; the major processes of transportation and deposition depend on climate just as much as different kinds of weathering and denudation produce different individual landforms and thus give a different aspect to the landscape's physiognomy.

Like tectonic processes, processes which modify the surface have been effective in all periods of the earth's history. Geology must study these processes during earlier periods with the same interest and care than it devotes to the present or more recent past. Geography, on the other hand, is not concerned with the processes of modification prior to decisive mountain building as such, only with their effects. Old degradational landforms, like old remnant surfaces or old deposits, are accepted as facts from geology. If we call them 'effaced' or 'fossil' landscapes and surface features, it clarifies the picture considerably. But even with reference to the more recent past, postdating the origin of the internal build and falling within the scope of geography, we can still speak of the landscape as developing, and indeed of its having a tectonic as well as a climatic development. Geomorphology, to follow the actualist approach in a logical fashion, must observe not only for how long processes have been effective but also the incursion of different past conditions. Thus landforms are of a diversity only partly susceptible to being treated as a whole, and partly amenable only to individual consideration.

That was roughly the position when Davis appeared as the prophet of a new theory which, as Rühl expressed it,[84] put an end to the dreadful and methodless geomorphology which had hitherto prevailed.

DAVIS'S 'CYCLE' THEORY

Davis called his 'method', by which he meant his theory, the method of structure, process, and stage. Since the terms 'structure' (*Struktur*) and 'process' (*Vorgang*) roughly correspond to the terms 'internal build' (*inneren Bau*) and 'surface modification' (*oberflachlichen Umbildung*) that I have previously used (although Davis used them in a rather narrower sense), his method is to be characterised by its stronger emphasis of stage (*Alter*). The method not only lays greater emphasis on the length of time for which processes have operated, than it does on the way processes differ in nature—with the result that it gives great prominence to the characterisation of landforms by age—but it also asserts that landforms have repeatedly come to the end of their life, only to begin it anew as the result of uplift—in other words, that landforms have usually developed through the course of several 'cycles'.

The question is whether this change of approach does in fact represent a step forward. It must be readily acknowledged that the careful tracing of how drainage patterns are subsequently modified, a fact not entirely unknown before, has thrown light on many hitherto obscure phenomena; but these phenomena have been seen as subsequent modifications when in fact they are the result of tectonic conditions of surviving (antecedent) river courses. The form of individual valleys and mountains does indeed depend to a certain extent on the length of time during which modifying forces have been effective; but time is far less important than the different ways in which modification takes place according to rock type and climate. To characterise landforms by their age leads to error. Whether old and senile valleys have ever been formed must remain an open question; Davis's 'old' valleys are, for the most part, valleys in weak rock; his senile valleys are either lowland valleys which have never had steeper or higher sides, or are not valleys at all but 'dells' (*dellen*), that is to say, smooth flat-floored hollows in benchlands or elevated surfaces. I will not dispute that commonplace fluvial degradation can bring about complete planation and the formation of a remnant surface—I developed the idea myself even before Davis did; but evidence that this has happened has not yet been produced. And it is questionable whether such an end-stage will be reached in nature. So-called senility can have quite different causes.

Thus the 'cycle' idea loses its real significance. That renewed uplift results in erosion being reinvigorated after periods of stillstand has long been known through the study of valley terraces; we have therefore long since spoken of periods of erosion. But a true cycle exists only when the life process is not broken in the stage of maturity or even youth, but progresses uninterrupted to the stage of senility, and only then, and not before, is reinvigorated. Such cases have not as yet been proven. The fact that geologically old remnant surfaces have been redissected in Tertiary and Quaternary times is something different.

Davis's purely geometrical approach is inseparable from its foundation on the deductive method. By deduction we can determine quantitatively the period during which processes have operated; but only observation will tell us the different kinds of process. While Davis talks a lot about 'life', his scheme lacks vitality, the landscape picture it gives has a moribund and dismal emptiness. The unending variety of rock types and the way they are arranged is submerged beneath the schematic contrast of 'hard' and 'weak'. Although he distinguishes a glacial and an arid cycle as well as the so-called 'normal' or fluvial one (in both of which his approach breaks down and deduction is seen to be an unusable tool), he considers the diversity of climatic influences in much too cursory a fashion. He takes no account of the differences in surface configuration within the fluvial cycle of polar regions, in the temperate zone, in steppe climates, in Mediterranean climate, or in the periodically and permanently moist tropics. Minor landforms have no place in the theory. It conveys nothing of the landscape's physiognomy or its structural style, and ignores many facets of its structural plan. Founded neither on rock type nor on climate, his description of the configuration of the earth's land surface is too superficial, and does not form part of the overall pattern of the earth's surface phenomena.

I can see Davis's approach only as an episode, not as a step forward in geomorphology. Its simplicity and the energy of its advocates has rapidly won for it a wide circle of adherents; it has enlivened geomorphological research and led to a number of correct results. But as a whole it has been abortive, and studies founded upon it have produced many fallacies. A lot of débris has to be cleared away to reopen the field to unrestricted research. As a whole, the earlier theory now condemned as backward was in fact on the right lines; it is upon this that we must build.

CHAPTER XIII
Landform Assemblages

CLASSIFICATION

We must now pass from our discussion of the origin of the land surface to a review of its landform assemblages. Each single surface feature is, strictly speaking, individual and different from all others. Provided that it is sufficiently important we are interested in each surface feature for its own sake and because it occurs where it does; we must represent it using all the resources of language, illustration, or maps at our disposal. But geomorphology cannot be content with this. It must progress from an appreciation of individual landforms to a treatment of landform types. This is necessary on formal grounds alone, since in this way repetitive and laborious description is avoided, and only in this way can phenomena of the same kind in an area be considered as a whole and compared with other phenomena. Moreover, only by singling out types do we obtain the indispensable foundation for induction based upon comparative study and the investigation of causes. But it is also of factual importance, since types correctly formulated express the relationships and similarities actually present, and point to identical or similar origins, to processes being of identical or similar effectiveness.

In geomorphology, as in any science, there are different levels of classification. A classification can be based on single characteristics: coasts can be classified on their groundplan, their elevation, or the type of rock of which they are made, and inland features on a single form characteristic. The classifications of older descriptive orography (*orographie*), such as we find in the first part of Songklar's *Orography* or in General Neuber's book* are largely such artificial classifications; only recently Passarge has provided us with another one.[85] They are not to be entirely passed over, since we have not yet got beyond this stage in our knowledge. But it is pedantry to force all geomorphological phenomena into the Procrustean bed of such artificial classifications. It is purposeless to set up landform groupings of features which only superficially resemble one another, as for example basins and talus-dammed lakes (*Aufschuttüngsseen*). Geomorphology must try to take an overall view of the properties of any phenomenon. It can do this by establishing types which are not *a priori* or intentionally genetic, but which take account of the phenomenon as a whole and are as a result of genetic significance; for if phenomena resemble one another in the overall nature of their characteristics this is always, or nearly always, because they are of similar origin. Many such types are in everyday use and have been adopted and further developed by geomorphology.

With conscious scientific intention F. Von Richthofen in particular used them in his classical *Führer für Forschungsreisende*. Inductively gained and initially descriptive, such types acquire genetical significance only as study proceeds. But as well as such types there are also landform classes deductively arrived at which have genetical implications inherent in them from the start; the validity of such implications must first be proven. What has already been said of deductive theories is true of these landform classes. They are trustworthy only when deduction follows painstaking induction, in other words is reproductive, or when it is subsequently substantiated by induction.

Every geographical classification must fulfil yet another requirement. It must simultaneously bring out the way in which phenomena are part of the nature of an area of the earth's surface, and show their relationship with other phenomena of the same landscape, as well as their arrangement and distribution. Only in this way can a classification be used both for regional characterisation and for the comparative approach of systematic geographical studies. Since a landscape depends both on internal build and on surface modification (factors which are unrelated to one another), geomorphological classification must also be multiple.

The landforms of the earth's surface vary greatly in size. From my window I look into the Neckar valley; I see the river filling the bottom of the valley, and the valley sides rising up out of the river bed; on the opposite bank the slope is cut by a tributary stream, higher up by gullies. I know that there in the woods there is a boulder field (*Felsenmeer*) and the valley side rises up to the Königstuhl. These landforms constitute a hierarchy: the Neckar valley is a major landform, its bottom and sides landform elements, with the boulder field and gullies as minor landforms giving variety to the landform elements of the slope. The only doubt I have is whether a side valley should be considered a major or a minor landform. The Königstuhl is a major one like the Neckar valley, but in contrast to the negative form of the valley it is a positive form. Neither the valley nor the mountain are independent features in any way conceivable in themselves; on the contrary, they are subdivisions of the Odenwald. In the Odenwald, and only at this level of the hierarchy, are we confronted by an independent surface feature, although it too, in its turn, is subordinate to a greater system, the mountain system of the Upper Rhine and the German Uplands as a whole.

To distinguish surface features of different rank is a valid approach for every upland. Although this will differ somewhat in highlands and lowlands, landforms of different size and rank always occur; they are of either a higher or lower order, and so cannot be compared with one another. It is very wrong to group together landforms of different rank on grounds of external similarity. The hierarchy can be compared with that of the human body, although the comparison must not be taken too far. An independent principle landform (*Hauptform*)* such as the Odenwald is like a man; individual dependent major and minor landforms are like such organs as his arm, his finger, or his fingernail; a group of mountains is like a family. Just as the finger cannot be compared with the whole man, so a single rock pillar (*Felspfeiler*) cannot be compared with a mountain; it is somewhat naïve to regard mountains as a collection of hills. Such a viewpoint is to be found in the work of Karl Ritter and has still not entirely disappeared. But even Fröbel disagreed with Ritter's view. He maintained that mountains are by origin primary and independent, that valleys and uplands, on the other hand, developed from them, and are therefore secondary and dependent. This view is nowadays so firmly established that geomorphology cannot but take it as its starting point.

Surface modification must not be seen as the same kind of causal factor as tectonic origin, as happens for example when eroded mountains are compared with block and fold mountains. Erosion is effective in all three, which each have a particular internal build. They differ because the original tectonic block from which eroded mountains were cut was a uniform, expressionless tableland, and that from which the other two kinds were formed were dissected from the outset. Even where aggradation takes place we must allow for the tectonic nature of the sub-surfaces, although it is usually incompletely known. Tectonic features alone are independent or autonomous; surface processes of degradation and deposition can only rework and reshape a tectonic feature, they can never produce something independent. Naturally, not every single tectonic feature, nor every individual basin, arching or fold, can be said to be independent, but only major folded masses. Even the majority of volcanic mountains are, like moraines or dunes, superimposed on mountains; only exceptionally are they independent.

Insofar as it seemed necessary for our purpose I have already reviewed the classification of dependent minor and major landforms. Now we must consider the geomorphological interpretation of the independent principal or aggregate features (*Gesamtformen*) of landscapes. Dependent major and minor landforms need not be brought in to the discussion except as properties or characteristics.

THE MORPHOLOGICAL CHARACTER OF LANDSCAPES

The two most important aspects of a landscape are its height and whether it is even or irregular; we can for this reason speak of lowlands, highlands and mountains. According to their absolute elevation above sea level and degree of unevenness we can distinguish further between uplands and hills, plains and lowlands, plateaux and highlands, etc. Uplands and mountains are of different types according to the kind of dissection to which they have been subject; his failure to study these differences is a major inadequacy of Davis's approach to landscape. The structural style or physiognomy of a landscape is a result of the kind of individual landforms which comprise it. On their own each of these aspects describes only part of the phenomena. It is therefore artificial to use any one as the basis for a classification of landforms; a natural classification must seek to take account of all of them.

Our discussion of geomorphological theories in preceding chapters must make it seem debatable whether the character of a region can be summed up in a single term, one from which a unitary classification of the earth's surface landforms might be developed. It seems to work in particular instances: when we speak of a volcanic cone, a simple fold mountain, an inselberg landscape (in its original sense), a morainic landscape, or a river plain, we visualise a distinctive unified picture. But then we unconsciously disregard certain important characteristics. In other cases we are at once conscious of an inherent duality. A landscape's morphological character depends both on its internal build and its past and present climate; since these are in no way related they cannot be included in a single term. All attempts at a monomial genetic classification must therefore prove unsatisfactory. If it is based on structure, it holds true only for fluvial landscapes, not glacial or desert ones; even then it is completely valid only for fluvial landscapes in moist climates. On the other hand, if we base it on surface modification it is valid only for a particular structure.

The geomorphological characterisation of any one landscape must be binomial, or in many instances threefold and multiple. A classification of landscapes is possible only by combining two, three, or even more criteria on which they may be grouped. A desert tableland, a moist tropical folded region, an uplifted remnant platform (*Rumpfplatte*) in the temperate zone, are such binomial characterisations. It seems that one can disregard internal build only in regions of deposition, where it is buried; but in fact levelness is a prerequisite even in this case, otherwise coherent deposition could not have taken place.

It will hardly be possible to speak of a landscape hierarchy based upon internal build and surface modification. Since surface configuration was at first seen as an almost direct expression of internal build, description and classification was based on internal build to the total exclusion of surface modification. Nowadays, many people are inclined to turn the tables; to go to this extreme is at best admissible only in regions where deposition predominates. But to use age as a principle on which to classify is even more one-sided, since it is far less important than either internal build or surface modification.

Since this book is concerned only with geomorphology in the stricter sense, *i.e.* with surface modification, and sees tectonics as inherent, here I will only briefly refer to tectonic description and classification. Again, internal build is a composite, not a simple condition. It is a matter of rock type as much as the way the rock is disposed. The tectonic surface, that is to say the surface which recent mountain-building processes would produce were it unmodified by exogenic forces, is the most important fact of all. In most regions structure is the outcome of events repeated over and over again. Though the oldest may be the most important in determining the future and therefore the most important for the geologist writing the history of the earth, the most recent major event is the one most relevant to the present and therefore to the geographer. To the geographer the German Uplands are primarily a block landscape, and only secondarily an old folded one, since this characteristic is now reflected only in the way the strata are disposed. General uplift and subsidence can certainly bring about the reinvigoration of erosion or deposition, or a change from one to the other; but they can only be incorporated into the characterisation as modifications.

For example, I would call the Black Forest a block oldland (*Rumpfschollengebirge*) partially covered by Bunter Sandstone. A more exact characterisation would include the extent and thickness of the Sandstone, the differing age and composition of the crystalline and sedimentary rocks which comprise the remnant mass, and the likelihood of its having been repeatedly uplifted; it would include in particular the amount of uplift that had taken place and the extent of the upland. If it should nevertheless prove true that the upland has been replaned in more recent geological time, this would also be incorporated into the characterisation. Our characterisation of the Swiss Jura or the Appeninnes will necessarily depend on whether we consider Brückner's claim that the former, and Braun's that the latter, have been planed off after folding, correct or not.* If this did not happen they would be described as simple fold mountains, in which case we would emphasise the predominantly limestone composition of the Jura, and the resultant karst. If they have been planed, then both mountain ranges would be considered similar to remnant oldlands (*Rumpfgebirge*), but of more recent origin. This would also be true of the Alps according to von Staff's interpretation;* but they seem to me and to others to be of a complicated structure, and to have been subsequently uplifted, perhaps warped, folded, and overthrusted. If one accepts more recent planation almost all tectonic differences disappear, or at least they become secondary; we are asked to accept that simple arching has occurred almost everywhere, and differs from place to place only in its magnitude.

But that is indeed pure imagination; provisionally we may and must hold fast to our classification of different kinds of tectonic features;

Landform Assemblages

we must distinguish between folded mountains of simple or complicated construction, dislocated fold mountains, blocks of different kinds (distinguishing them firstly by the location and form of the whole block, and then according to their internal build) and finally volcanic mountains. Still other types will probably come to light, types we cannot now properly assess; for the difference between mountains such as the Tienshan and the German Uplands cannot simply depend on differences of scale. This would be the case if, as is now held, both are remnant oldlands; but the differences between them must be related to their essentially different processes of formation.

CLASSIFICATION OF SURFACE MODIFICATION

Such a tectonic description and classification works within any one type of climate; for then the processes of surface modification depend on internal build, and the landforms produced can be considered to be functions of structure. But where climates differ, surface modification differs even if the structure remains the same. Moreover, we must also consider the climate of the past. Imagine the Odenwald, the Harz, or any of the other Central German Uplands put in turn into different climatic zones and try to imagine the landforms they would then take on! Even in Scotland and Norway, and even more so in higher latitudes, the Pleistocene snowline would have been so much lower and glaciers descend to such low altitudes that glacial would outweigh fluvial configuration. The upland would have a glacial landform assemblage. If it were moved into the Sahara there would be no regular drainage, dissection would be slight, and chemical decomposition absent or of a different kind; the upland would have the landform assemblage of a desert. In the moist tropics it would be dissected as in Germany; but the different kind of weathering and denudation in the tropics would shape different landforms. In the periodically-humid tropics the upland would probably become an inselberg landscape. If the differences between climates were less, the less striking would be their different effect on the upland's form.

As climate changes, so do the processes by which material is relocated, such as weathering and denudation, and major and minor landforms. Major landforms are the more durable, and have often been formed in the past when a different climate from that of today prevailed; minor ones belong predominately to the geological present.

Moreover, weathering and denudation are directly related to conditions on a particular part of the earth; the major processes of transportation and redeposition, on the other hand, extend their effect into neighbouring regions, so that, for example, an area that has never had a permanent covering of snow can nevertheless be glaciated and display a glacial surface configuration; the rivers of a humid area may enter an adjoining arid area.

These processes of transportation and redeposition are of paramount importance in interpreting and classifying landforms. We must therefore begin by broadening our characterisation based upon internal build to include the position a landscape occupies in a régime of material being moved from one place to another. In doing so we must distinguish on the one hand between the different factors in the relocation of material, on the other between regions of degradation and deposition. In this way the major traits of surface modification are represented. But the physiognomy of mountains emerges completely only when we include present-day changes in our description.

The Lake District is a block oldland (*Rumpfschollengebirge*) and in this respect therefore corresponds to the Central German Uplands. But unlike the Uplands, the Lake District was for the most part glaciated during the Pleistocene, and is now subject to a moister climate; the Ahaggar Plateau is a limestone desert tableland; and many other examples could be cited.

Although differences due to internal build must be fundamentally distinguished from those of surface modification brought about by climate, the particular characteristics of internal build can nevertheless have so marked an effect that differences of climate become secondary. Their effect can even be so great as to produce landforms which would in other circumstances be seen as belonging to another kind of climate. We can therefore speak of convergent phenomena. While this is the case especially in limestone or karst mountains, it is also true, for example, of permeable sandstones like the Quadersandstein of Saxony, Silesia and Bohemia. The landforms of these areas are so much like those of desert areas that they were wrongly explained as the result of a desert climate in the past. Fault scarps resemble the escarpments formed by degradation that we call cuestas. It seems that rock platforms can be formed by marine abrasion as well as by sub-aerial degradation of various kinds.

CHAPTER XIV
Subdivision and Grouping of the Land Surface

The subdivision and grouping of the land surface is something rather different from the classification of surface features. While the classification of surface features involves a comparative review and interpretation of landforms by types, without reference to their spatial association with one another, the subdivision and grouping of the land surface is concerned precisely with their spatial associations. We ask such questions as, which parts of the land surface should be grouped together and which separated, and where should the boundary lines necessary to any clear treatment be drawn? In doing so we are only concerned here with the earth's land surface as such; in a systematic geographical treatment we would always have to ask ourselves whether we should attempt such a geomorphological grouping at all, or whether it would not be better to base a grouping on drainage, climate, the plant-cover, or human phenomena.

The earth's land surface is grouped primarily on a tectonic, rather than a geomorphological, basis. Not only the contrasts between continents, but the distribution of mountains, highlands and lowlands is also for the most part inherent in the internal build of the earth's land surface. Tectonic processes alone produce independent principal landforms. The processes of surface modification, in other words geomorphological processes in the strict sense, need not be considered until we come to subdivide independent principal landforms; even then they do not work alone but in conjunction and competition with internal build. We must often be arbitrary, expedient and tactful in giving precedence to one or the other.

But quite apart from this we can approach the task of grouping from two different viewpoints: we can lay the emphasis mainly on the type and nature of surface configuration on landform assemblages, or on areal association as against areal separation. Differences of internal build, which are usually of the greatest and most impressive kind, can give rise to different kinds of surface configuration and landform assemblages. The Alps and their Foreland, the Alps and the Swiss Jura, the crystalline Central Alps and the Limestone Alps, the folded and tabular Jura, the Black Forest and the Upper Rhine trench, the Elbe Sandstone Mountains and the Lausitz Plateau or the Erzegebirge, are contrasts based upon internal build. On the other hand, when adjoining uplands are related in origin they can be treated as one mountain system; the Alps with the Carpathians, the Dinaric Alps with the Pindus, or the German Uplands as a whole. Even where the structure is the same, differences of rock type can lead to significant differences in surface features, and so influence the plant world, and man himself. This is especially so in scarplands, where breaks of slope are on the whole coincidental with changes in rock type. How different the Muschelkalk plateau is from the Bunter Sandstone in the scarplands of South West Germany! It is so different that the boundary between them is seen as the boundary between the Black Forest, the Odenwald and the Spessart and neighbouring areas to the east; similar changes occur where the Muschelkalk meets the Keuper, and the Lower and Middle Jurassic meet the Upper Jurassic.

Landforms also vary from climate to climate. This variation can take place gradually when an area of uniform structure extends across different climate zones; but it can also be abrupt when a high mountain ridge forms a climatic divide such as that between the Indian and Tibetan sides of the Himalayas, or the eastern slope of the Peruvian–Bolivian Eastern Cordillera and the Sierra and Titicaca Highland. Areal separation is especially prominent when protuberances adjoin depressions. When this is the case one can focus attention on either the protuberances, the positive landforms, or on the depressions, the negative landforms. The geomorphologist will usually prefer to concentrate on the former, the human geographer on the latter. Negative landforms may be tectonic features, for example trenches and basins, even perhaps just the space between positive landforms independently formed; but they can also be erosional features such as valleys and valley-like depressions. When they are the former, mountain valleys have a measure of tectonic independence; in the latter case, they are units carved out of the landscape, but none the less clearly demarcated from their surroundings. The older school of orography preferred to subdivide the Alps on the basis of areal separation rather than on tectonic or geomorphological individuality. Although the tectonic and geomorphological individuality viewpoint must now be seen as the more correct one, it is wrong from a geographical point of view to give the tectonic and geomorphological approach a higher scientific status than areal separation, as some geologists have.

Appendix: Geomorphological Research and Presentation

METHODS OF GEOMORPHOLOGICAL RESEARCH: COMPARATIVE MAP STUDY

A dispute as to method was a feature even of the birth of modern geomorphology. Peschel's *Neue Probleme der Vergleichenden Erdkunde*, the importance of which, in introducing the causal treatment of geomorphological problems into geography, cannot be denied, traced the distribution of particular surface features like fjords, valleys, lakes, and deltas over the earth's surface, and by comparing their distribution with that of other phenomena, reached conclusions as to their causes. This way of working contrasts with that of geology, which usually studies individual phenomena. Certainly, a world-wide comparative approach was not entirely new; but in this case it was applied to a whole series of problems, recognised as such, and was handled with such elegance that it carried people away. Unquestionably, this approach gave us a number of important insights. But other results so gained did not stand up to test; gradually the method's failings, which individual sceptics were quick to point out, became more and more evident. Peschel's world-wide comparisons were founded on the study of small-scale maps and individual travelogues, and were, as a result, inherently superficial. But Rudolf Credner's* and other research workers' penetrating investigations of deltas, relict lakes, and other phenomena also had a large element of uncertainty because the cartographic and literary material on which they drew had not been collected with the particular investigation in mind, and so did not provide an adequate foundation.

I recall a dramatic scene at the Geographical Congress in Halle. After Credner had delivered a polished lecture on the Alpine lakes using Peschel's method, Zittel rose and communicated Penck's recently-completed investigations on the role of glacial erosion in the origin of the Bavarian Alpine Foreland. The superiority of observation over comparative map study was evident to all. And although Penck's conclusions have since been disputed, the argument is carried on on the basis of observed facts not map study.

Our most important insights into not only minor landforms, which have to be directly observed (illustrations provide only a limited substitute), but also the erosional nature of valleys, glacierisation, and the surface configuration of deserts, the tropics, etc., has been obtained by direct observation. Of course we must not go to the other extreme and ban map study altogether. To study a phenomenon by direct observation is always a laborious undertaking, and is possible only to a limited extent; if we were to limit ourselves to direct observation alone we would never reach general propositions. It must be supported by the comparative study of maps. Such study draws attention to problems, and even to certain possible solutions. When observation has provided us with a result and the picture is brought into focus, we can hazard more sweeping conclusions on the basis of maps and literature. But knowledge gained in this way remains provisional; it is not established until observation has shown it to be valid for the whole earth. Modern geomorphology still depends too much on knowledge of small areas. The earliest work was done in West and Central Europe, followed by work in the North American Cordillera. Richthofen, in particular, carried penetrating research to Asia, other travellers following his example in different parts of the world. But they had to carry out their work in a pioneer fashion at first, extending their research over wide areas, producing topographical and geological maps for themselves, before they could begin to tackle geomorphological problems. Their work cannot have the same accuracy as detailed European research. We must have more such journeys and researches, and they must be continually renewed; new problems constantly emerge and methods of study become refined. Only in this way can we overcome the narrowness which still besets modern geomorphology, and gain new viewpoints and results of general validity.

METHODS OF GEOMORPHOLOGICAL RESEARCH: INDUCTION AND DEDUCTION

Though vital in other branches of natural science, experiment still plays a rather small part in geomorphology. Some of the well-known experiments of Daubrée* and other particular geological experiments might be mentioned here. The work of running water is studied on a large scale in hydrological laboratories. In the main we have to depend on normal induction and deduction.

The inductive method in no way entirely excludes conclusions being arrived at deductively; on the contrary, it just cannot do without them. But such conclusions are in general made to follow upon those reached inductively. Sometimes deductive inferences even outrun inductive, though this is contrary to a strictly methodical approach. Work as early as that of Powell was markedly deductive, and in the fifth chapter of his celebrated book on the Henry Mountains, Gilbert presented a comprehensive deductive theory of erosion and degradation. There are some deductive elements in Richthofen's loess theory, still more in his theory of abrasion. His *Führer für Forschungsreisende* is full of deductive conclusions. Philippson's contribution* to the theory of erosion, arising out of his study of watersheds, is deductive. I myself, in my studies of Saxon Switzerland and elsewhere, used deductive methods a lot, in fact sometimes too much.[86] Passarge, too, made many deductions, and many other examples might be cited. But we can characterise such use of the deductive method as a whole as being reproductive; it establishes the ways in which phenomena are causally associated.

The theory of erosion was developed from the laws governing the movement of flowing water, only after it had been established that valleys were the result of erosion. And even today, after the theory has been developed, each individual study begins with a diagnosis based upon observation. Before we consider an elongated depression as an erosional feature, we study it to see if it has in fact the characteristics of a true valley. Direct evidence for this is sought in valley terraces, valley windings, etc. Before a scarp is explained as a feature of sub-aerial degradation we must exclude that it was formed by faulting or as the result of wave attack. We must establish that there is no evidence for such an origin, and on the other hand show that the sequence of rock types provides conditions suited to the formation of an escarpment.

Davis's older works, which appeared some years later than the German and French geomorphological works we have mentioned, essentially followed the same method. In his book *Die Erklärende Beschreibung der Landformen** he sketches the conventional inductive method as the method of geomorphological research. But he

Appendix

does not seem to regard this as induction, since there are deductive comments in it as well. Much of the deduction in this, as in other summary texts, only appears to be new research. In fact it is only a representation of knowledge previously gained by induction in a deductive guise. Gradually, perhaps without realising it and without wishing it, Davis slipped into deduction in his research as well, and led his followers along this path. Today, the research method of his school, to which, in this respect, Walther Penck may also be reckoned to belong, is deductive. When Davis protests, in a review of the present work's first edition, that this is not so, since every assertion has been subsequently examined in the light of the facts, he is merely showing that he cannot distinguish between the two methods.

Davis's method is deductive in that it first derives landforms from the forces which are effective, and only afterwards compares them with reality. The scheme is complete before it is confronted and filled out by reality. At the International Geographical Congress in Geneva, Davis proclaimed emphatically: 'The geomorphologist should first go into a darkened room and closing his eyes think, think, think; only then should he go out into the field.' The same idea repeatedly recurs in his book and dominates the working method of his school. When a follower of Davis goes out into the field or studies the map, he delivers a genetical judgement at the first glance, translating observation into preconceived terms. Inductive research begins with a description involving nothing but the facts, and progresses to a causal interpretation. Genetical concepts are reached by gradual refinement and clarification of initially purely empirical terms. Observations stand even when theory changes. The deductive method, on the other hand, first conceives the genetical concept and then takes it into the field. As Rühl puts it,[87] the research worker can proceed quite systematically (for 'systematically' he should really have said 'schematically') in his field investigations. 'When once this work has been completed we shall be able to incorporate every surface feature with which we are confronted in nature, into one series or another.' Conclusions are swiftly reached this way; the only question is whether they are worth anything.

Clarity and exactness in interpretation and terminology is the primary need of every deductive theory, and it is precisely this that Davis's theory entirely lacks. Its emphasis on comparing landscape development and life is no explanation; even its terms are questionable, vacillating between the length of time during which a landscape has been developed and its stage of development. But the expression 'stage of development', now in favour, is not a simple straightforward term; it contains two ideas, namely duration and intensity of development; in other words, it is initially only a descriptive, not an explanatory term. Davis's second major term is just as inadequate. The term 'cycle' does not distinguish between a very short and a very long interruption in erosion, and makes no allowance at all for subsidence or other interruption of erosional processes. Nor are the other principal terms, 'structure' and 'process', sharply defined. It is not clear whether they refer to external (exogenic) processes only or to internal (endogenic) ones as well. Secondly, Davis cites texture and relief as conditioning the shape of the earth's surface; but he does not indicate what relationship these have to the three determining factors already named, nor whether they are derived from them or are in fact independent.

In essence, Davis's theory amounts to saying that in the course of time valleys are deepened and their sides flattened; he concedes that differences in rock resistance exert a certain influence on the speed with which processes work. It is a purely geometrical construction, and not a causal explanation based on the physical and chemical nature of the processes. Only glacial and arid cycles are separately considered, since running water then plays little or no part. On the other hand, no account is taken of the diversity of processes according to rock type and climate in the normal or fluvial cycle; the manifold variety of landforms which arises from these differences goes unmastered. Since Davis renounces an actual explanation based on processes he gets his results with an ease that does not match reality. There is nothing simpler than transforming a valley into a peneplain by geometrical construction, by gradually flattening the valley sides; but in nature this is a phenomenal process requiring an immense time span. It is true that it is theoretically possible, but it is doubtful whether it actually takes place, at least in our European climate. For this reason we cannot postulate peneplains without compelling reason, just because they are a convenient aid to applying a scheme of erosion. In research one must always begin by making the simplest assumption; but this is the one that taxes nature least, not the one that makes for the easiest explanation.

Appendix

From its nature, deduction can only point to possibilities and show that a landform can have come about in a certain way; but the same landform may be formed by other processes as well. Only an exact analysis of processes, a 'differential diagnosis' as Passarge says, can decide whether this is so or not. But most of the works of Davis's school fail to do this,[88] and in fact most landforms can be better explained in another way. Peneplains are usually unnecessary. Geographers and geologists influenced by Davis's method and theory have been in a delirium like that which followed the publication of Peschel's *Neue Probleme der Vergleichende Erdkunde*. The elegance of the deductions, and the ease with which apparent results are gained, compared with the slowness of sound induction, together with Davis's unquestionably great didactic and artistic skill, make this delirium psychologically understandable. But the hangover cannot continue. The path of science is both long and thorny; castles in the air do not survive.

One way of preserving the deductive approach would be to blame the master; but when an important man devotes his entire energy to some end and fails, it is a bad sign. Deduction is a reasonably certain way to knowledge only in the abstract sciences, such as mathematical physics, where the number of causes and conditions under consideration are small and the results deduced quantitatively, in other words attainable with the aid of mathematics. Even in chemistry it can play only a much smaller role; and in the concrete experimental sciences in which each single case involves unknown causes and conditions, and quantitative treatment is only exceptionally possible, it is of little importance. Then it must be combined with the inductive approach, and can only be an aid to clarifying ideas and stimulating keener observation and diagnosis. Particular concepts based upon deductive theories will survive as part of a geomorphology of the earth's surface; but the edifice will collapse. Only sound inductive research leads to the goal.

METHODS OF PRESENTATION

We need to study not only geomorphological research but also geomorphological presentation, especially since this had a prominent place in Davis's approach. Indeed, his whole doctrine culminated in a special form of presentation which he called explanatory description. But in this study I will not, as Davis did, discuss purely didactic questions as to how the lecturer can put himself into a favourable light before his hearers. I intend to treat the problem as one of scientific theory. Naturally, I am not saying that appearances can be neglected; but they can be secondary. Logic is the main thing.

Depending on the state of geomorphological knowledge and the purpose for which a presentation is to be used, it has to be either a description in the strict sense of the word, and aimed at the facts, or an explanation aimed at causes. Davis believed that when he applied his closed, deductively-reached schema to reality he could immediately explain this reality, that is to say, could describe it genetically. But the validity of his procedure stands or falls according to how applicable the deductive method of research is. If such a method led to reliable conclusions, then to give a description which was from the outset explanatory would also be justified. But faced with the diverse and phenomenal reality that is the form of the land surface, it fails. It feigns a knowledge we do not have and therefore makes geomorphology untrustworthy.

All the forms of the earth's land surface must first be treated as facts and brought under review. This step is often foregone by those who exaggerate causality. It is a well-known fact that the student who has not been through a strict schooling usually answers the question 'how' by an explanation of 'why'. This same failing is widespread in geomorphology, and is raised to the standing of a principle by Davis's school. One can read long monographs and books without having any idea what the features actually look like. Description, that is the recounting of the facts, can come either at the beginning of an analytical account, or at the end of a synthetical one, showing the facts of the case and clinching the argument; but it must always be there. It must not be fragmented into details and secondary matter as is often the case in older geographical literature and the works of laymen. It must give the essentials and be lucid so that the land's shape stands out clearly in the mind's eye, and so that the less easily-seen, yet significant, facts can also be clearly conceived. The methods of scientific expression must be continually re-examined to see if they measure up to these demands. This does not seem to me to be the case with description according to Davis's method. Not only is the structural style given to a landscape by its minor landforms not apparent, but it is clear only in the simpler cases. Designation by age is not without a certain graphical force to be sure; it has only the minor failing that it is wrong. But hypothetical peneplains and imaginary cycles do away with perspicuity. Does one get a clear impression of, say, the German scarplands from Braun's book on Germany, a book that can be seen as the test-piece in regional geography of Davis's method of presentation?

Explanation, the demonstration of causes, is only the second objective. All landforms have developed. Most minor landforms probably belong to the present; but major forms date from the geological past and display tectonic and climatic development. Now any development can be most easily represented by historical narrative. It is rightly the method of presentation used by historical geology, for it shows how the different endogenic and exogenic processes are related or change with time, and consequently how development comes about. When, several decades ago, geography changed over from purely descriptive to explanatory or genetical interpretation, geographers had often seen the genetical method used in geological narrative, and they used the geological description of landscape development in their regional descriptions. Many geographers find the genetical interpretation of peoples, states, settlement, etc., impossible without an historical narrative. But in this way the boundary of the two sciences has become obscured. Quite apart from the fact that geography must as a result adopt the results of geological research, and that our knowledge of geological development is still in many cases very uncertain, and that in this way a lot of unnecessary uncertainty is introduced into geographical accounts, development as such is of no geographical interest. It is immaterial to the interpretation of present-day conditions whether at a given place slates are Devonian or Cambrian, or whether the Bunter and the Keuper Sandstones are of marine or sub-aerial origin. The only fact of geographical importance is that a thick strata complex of a particular make-up exists. Davis[89] and I are in agreement on the methodological principle that geography must not end up as a geological narrative. But beyond this our ways diverge. Davis believes himself able to describe landforms in terms of their age and the cycle; I think that he is deluding himself because his theory of stage and cycle is faulty. Genetical interpretation without geological narrative must take internal build as its starting point; but it must accept this as a given fact and introduce the causal approach to account for its modification by surface forces.

Every description must work in terms of specific types, that is, with a classification of phenomena. If the specific types of most elementary descriptions are artificial, based upon a single characteristic, they are later replaced by types which, though still descriptive, embrace several characteristics. In this way the foundation for a genetic interpretation is built up, and results finally in a truly genetic terminology based on the diversity of formative processes.

In the geographical study of landforms we must always take care that genetic terms are based on properties characteristic of different lands and landscapes. It must develop out of the internal build of an area or out of its climate, together with its drainage and its plant-cover. The surface features must be linked with one or other of these conditioning factors. Only thus can all the landforms of an area be united in an assemblage distinctive of it and distinguishing it from other areas. Moreover, only then will the classification of surface features be the means by which we can gain a comparative appreciation of their world distribution.

TERMINOLOGY

Every science needs technical expressions with the help of which it can designate phenomena briefly yet clearly. The marked need for these in geomorphology today is a welcome indication of its virile scientific development. But the importance of terms can also be exaggerated. They are an aid, a way of conveying information in an abbreviated time-saving way. A technical expression should convey in a single or few words a concept which would otherwise have to be laboriously described. But to be of any use it must be familiar; unusual technical expressions are a hindrance rather than an aid.

Here I am not speaking of the need for a terminology in all branches of geography[90] and in science as a whole. Nor am I speaking of the often excessive consideration for priority of naming, international or national intelligibility, linguistic correctness (although it is precisely in geomorphology that one sometimes wishes for this), or of clarity and purity. I am concerned only with the viewpoint distinctive of geomorphology.

At one time geomorphological terms often acquired a hydrological

Appendix 157

nuance, because surface features were regarded as static things on which river courses and watersheds were established; the entire course of a river was termed its valley, the whole of its area, its basin. In so doing these terms lost much of their value. The word 'pass' was given both a geomorphological meaning and a meaning familiar to students of the geography of communications. Thus the word has been applied to both a saddle or a col in a ridge, unused by any routeway, or a pass which is not in fact a saddle but a defile, used by a road. The term 'pass' should either be left entirely to geomorphology, or more appropriately to the geography of communications.

More recently, geomorphological and tectonic terms and expressions have been confused with each other. Geomorphology must use tectonic terms because surface features depend to a high degree on internal build; but geomorphologists must remember that only in the rarest cases does internal build remain intact. Much more often it is markedly modified by surface processes. For this reason tectonic expressions cannot completely convey form. On the other hand, geologists have adopted from mining terms like trough, saddle, arch, trench, horst, etc., to designate internal build. And such terms were used correctly at first. But time has brought inconsistencies. Tectonic arches or upfolds often become valleys, and conversely tectonic troughs become ridges. Horsts can be depressions, trenches protuberances. These terms should be limited to the instances where exterior form reflects the internal build, as when a tectonic trench still actually is a trench-like depression. For purely tectonic use we should select other terms such as the Greek anticline, syncline and so on. Or we should use such terms as strata-depression. It will be inconvenient and therefore difficult to discard all at once old expressions which have become deeply rooted: but we should take care when creating new ones. It is absurd for Brancá* to call tuff hills (*Tuffberge*) maars because they have resulted from the modification of maars.

The terms commonly used for landforms are of various origins. A number of them are drawn from everyday life: many surface features are so prominent that they play an important role in our life, and a special word was soon coined for them. In other cases, the need to give them a term was not felt until much later; instead of being given a special term they were grouped along with features of a similar shape. In this way expressions such as furrow, basin, pan, trough, trench, horst, saddle, funnel-shaped valley-mouth, delta, and many others came to be used in geomorphology. In still other instances a name originally used in only one locality, in Germany or elsewhere, was taken into scientific terminology. Such expressions as haff, klamm, fjord, ria, liman, canyon, caldera have long since established themselves; playa, bolson, cuesta, and many others have recently been adopted. In particular cases a general meaning has been given to individual names. One of the best known examples of this is the word meander; another is monadnock, the name which Davis gave to a particular form of hill after an example in New England. We must be sparing with both kinds of name, for they are a considerable and often useless burden on the memory. It is especially annoying when these new expressions, such as the Spanish ones mentioned, do not correspond with their true meaning: for instance, *playa* means shore, *cuesta* means slope; we must know the language from which the word is taken.

All these terms refer in the first instance to external appearance. Though they serve to describe the features, they tell us nothing about the features' origin, and most of them are also inexact. Since geomorphology needs a strict and unequivocal terminology, it must give these terms more exactness; it must therefore widen or narrow their meaning. Many geographers, especially military geographers like Neuber, believe that for all that terms must never depart from being purely descriptive, that orographical terms must not have a genetical meaning. But the general terms of a natural classification acquire of themselves a genetical meaning as well. Thus in modifying terms we should always aim to define expressions in such a way that they not only denote a particular landform in unequivocal fashion, but also correspond to a particular origin. Of course, different formative processes may produce very similar, convergent landforms; whether these forms are in fact identical is a matter for debate. In our research work we must seek to recognise those characteristics which correspond to different origins, and incorporate these into the description and designation of landforms.

While it is generally true that terminology like a new definition begins by being descriptive, and only later acquires a genetic meaning, it can nevertheless be based upon origin from the start. Powell's distinction between 'antecedent' and 'consequent' valleys, for example, and Spethmann's use of the term *Hartlinge* for isolated hills thought to have resisted destruction because of the hardness of

Appendix

their rock, are so based. It cannot be said that all such expressions have stood up to test. The genetical terminology of Davis and his school using the deductive approach was the most systematically developed but it has collapsed along with his approach.

Like classification, geomorphology's genetical terminology must allow for the interaction of internal build and surface modification, and must therefore always be binomial or multiple. As far as geography is concerned, structural terms need only be descriptive even when the state of geological knowledge makes genetical ones possible. It is of no consequence to geography, since it is uninterested in the question, how the landforms of the geological past, for example the remnant surfaces of the Permo-Carboniferous period, originated. In most cases they have had a complex developmental history; our terminology must take account of the nature of the last major dislocation just as much as the outcome of what happened earlier. So it is not enough for us to designate a piece of the earth's crust as a block; we must go on to say whether it is a tabular, remnant or folded block, and particularise its rock composition. On the other hand, the modifying processes call for a genetical terminology based primarily on their place in the major systems of material relocation, secondarily on the kind of weathering and denudation, and only lastly on age, that is on the time during which surface modification has been effective.

It is scarcely possible or expedient to coin a special expression for the landforms characteristic of every process, as Passarge does. They easily become monstrous words. It is better to give a description which includes what is essential in brief statements. Most of Swabia is a mildly inclined, little faulted tableland which, thanks to the marked differences of rock type, has been transformed into a scarpland. An upland such as the Odenwald can be characterised as a weakly inclined transgression block, which, as the result of intense degradation in its front part, has been divided into a folded remnant and a sandstone tableland; both parts have been transformed by river erosion into a valley landscape. In the case of the Black Forest the greater altitude and the glacial modification of the higher parts would be stressed. Age does not seem to play any critical role in these uplands; it is to be brought out only when valley terraces become more important and are associated with major planations. Past climate and relationship to the sea can also come into the short description. The west coast of Norway is the marginal slope of an old uplifted block remnant, probably repeatedly glaciated along river-cut valleys, which were then submerged beneath sea level to become fjords.

Further examples are unnecessary. We need only point out here how with this kind of terminology one can arrive at an expression which embraces a landscape's principal characteristics, an expression which on the one hand links up with internal build, on the other with regional climatic differences of both today and the past. Moreover, it can from the outset be an expression conveying the geographical distribution of surface features; expressions based upon age, or purely descriptive ones, convey nothing of this. A comprehensive classification of surface features is beyond the scope of our discussion here.

OROMETRY

Geomorphological descriptions should, whenever possible, particularise the altitude, relief energy and slope. One of the principal aims of the map is to portray these characteristics by means of contour lines (isohypses), hachuring or shading. But in verbal description, on the other hand, to use too many numerical values is clumsy; only a few selected, and particularly characteristic, values should be given. But it is more important to enquire whether we cannot express the entire character of a landscape numerically. Ideas used in analytical geometry to express geometrical figures arithmetically have been copied in orometry and paralometry (or, in a word, morphometry). But these attempts have sought to solve the problem not by curve analysis, which would be impossible since all the outlines of uplands are irregular, but by using average values, such as the mean height of summits, passes, crestlines, of entire mountains or mountain groups, as well as the average gradient.

The aim of doing this is to bring out landforms clearly and free them of fortuitous aberrations in order to be able to undertake a comparative treatment. The idea stems from Humboldt,* who calculated the average height of the continents. Songklar* and others aimed at more than this, and developed methods of calculation with great acumen. But they have certainly taken the significance of average values as too self-evident. The average height of continents or larger mountains and mountain fragments serves us to calculate their volume; it there-

Appendix

fore has geophysical and geotectonic importance. But what does the mean height of ridges, summits, or passes, or the average gradient, tell us? They occur nowhere in nature, nor are they ideal values by comparison with which actual values are anomalies. We cannot conceive of any force which has worked to produce average values. Genetical study of the earth's land surface cannot begin with averages. I have never found that they have any role to play in this task. Nor have they any more value for the geographical study of settlement and communications. Man does not live at an average altitude, but in valleys and on mountains. The actual, not the average, height of passes is of importance for routeways; they can just as readily use passes at quite different altitudes as at roughly the same altitudes.

MORPHOLOGICAL MAPS

While geological maps are the indispensable foundation for geomorphological studies, it may nevertheless be advisable to appraise individually such particular rock properties as hardness or compactness, permeability, solubility, liability to disintegrate, etc., as well as the influences of other natural realms, for example vegetation, and show their distribution. Passarge in particular has shown the value of typical examples of such maps; he provided one such example with reference to the Stadtremda sheet (in Thuringia, south of Weimar);* later, similar maps were drawn for trench warfare. Unfortunately, Passarge has said nothing about how they should be drawn. In most instances it will hardly be possible to learn about the properties named by direct observation over the whole of an area; we shall be able to determine what they are like only along particular transects and then have to extend our knowledge areally with the help of geological maps. But then there is the ever-present risk of overlooking small but significant differences of rock facies, and so wrongly assess the properties of a locality; I know that this happened during the war. Owing to their high cost few such maps can be drawn, or at least published; but sample studies of this sort are certainly to be recommended as a basis for further geomorphological studies.

Furthermore, the processes of surface modification can be portrayed cartographically. Naturally, this would be done only for those processes directly observed, not those inferred from landforms or soils. Minor processes of weathering and mass wasting, for example sheet-wash and the rinsing out of materials, as well as such major processes as rock falls, landslides and the like, or the work of water at times of maximum and minimum flow, can be observed by the methods described by Götzinger,* and their distribution mapped directly. Such maps can just as well be drawn at a large scale for small areas, even if only on a selective basis, as at a small scale for whole countries or even the whole earth. Soil maps can also be used for comparative purposes, since soil development is closely linked with the shaping of landforms; such maps can tell us much about the rocks, the climate, and the type of processes. Passarge was right to include a soil map in his geomorphological atlas.

To a certain extent the common topographical map showing relief by contour lines, hachures (or shading) and additional symbols, is a map of landforms. But although the most recent maps, particularly the fine maps of the Austro-German Alpine Society, are less schematically drawn than most older ones, and seek to show landforms as well as the height and slope of the land, landform type is inadequately brought out.

Special morphological maps[91] which aim to show landforms as such are therefore justified along with the usual topographical ones. They can be drawn at the most varied scales; but they must be based on observation, and portray only observed, if conceptualised, facts.

The distribution of such distinctive landforms as earth pyramids, caves, rock pillars, blockfields, corries and rock basins, terraces, dells, gullies and so on, can be shown on special morphological maps plotted on either large-scale plans or small-scale maps; such maps will be of great help in analytical study. But a more important aim is for them to portray the landform assemblage of an area as a whole, when different but associated landforms are collectively represented by a single symbol. This is the conceptual process of setting up types, applied to the overall character of a landscape. It involves the portrayal of the landform assemblage, for this determines the structural style, as well as slope conditions. The maps of Gehne, Behrmann, Passarge, Machatschek, Rathjens,* and many others, aim at this; but they sometimes represent theories rather than observed landforms. Transects are an adequate substitute for maps, particularly in areas of tabular strata.

Even more generalisation is called for on small-scale morphological maps. These show whole areas, each part with its own landform assemblage side by side. As important as such maps are, and as frequently as they have already been used to illustrate works of regional geography, they can show only the most generalised landform types. They show escarpments and planations, or mountain ranges of a particular kind more clearly than the usual topographical map. If the area shows tectonic as well as climatic differences, a twofold method of representation would be used; differences in the kind of surface modification might be shown by colours, differences in the tectonic foundation by hachures.

Machatschek contrasts those maps which, even though generalised, portray facts based upon direct observation and mental reflection, supplemented by constructs and grouped into types, with those which display the geomorphological viewpoint of their author. The maps which he cites as examples of the latter are largely those which seek to distinguish stages of development (*Altersstufen*) and single out peneplains of different ages; in other words, they are cartographical representations of subjective genetical concepts based upon Davis's theories. Naturally, maps of the same character, as far as their method of representation is concerned, can be drawn from different theoretical standpoints, as for example when Passarge records certain soils as periglacial. Such maps cannot be rejected on principle; for mapmaking impels one towards more complete observation, and provides the basis for comparative study. But we must exercise restraint in publishing them. One should not forget that because of the positiveness with which it represents features, a map is readily taken to be objectively correct, and the hypothetical in it easily forgotten. It is permissible to go further in advancing hypothetical interpretations in a specialised study intended for professionals than in one intended for a wider, uncritical circle of readers; a map like Braun's small-scale geomorphological map of Germany must do a certain amount of harm.

ILLUSTRATIONS AND VIEWS

Photography is the principal way in which we now obtain perspective views. Of course only the professional can usually so select the subject and viewpoint as to give the photograph scientific value. But all photographs, since they directly reproduce reality, suffer from having too much non-essential detail. What they are intended to show is often obscured by extraneous things like vegetation or dwellings. For this reason sketches have not lost their value; indeed they have long been far too neglected. Scientific-geomorphological sketches of great perfection are to be found in the publications of the *Geological Survey* on the Cordillera of the American West. The services of professional draughtsmen were used, and the bareness of the ground favoured the anatomical portrayal of relief. Because of the difficulty of making them, Alphons Stübel's drawings of the volcanoes of Equador and Columbia are even more astonishing.*

Hayden's companion Holmes, to whom we are indebted for especially fine views, was in fact the inventor of the block diagram, in which square or rectangular segments of the landscape are drawn and the geological structure added as a profile on the front and side.[92] Later, Davis adopted and popularised these block diagrams. But he did more than this. He not only drew block diagrams of actual localities, but also and, in far greater numbers, schematic block diagrams of landscapes as he thought they originated according to his theoretical views. The same is true of these diagrams as of morphological maps drawn on the basis of genetical concepts. The masterly skill with which they are drawn makes them persuasive; more than anything else they have perhaps served to disseminate his teaching widely. Davis holds that beginners can learn to understand geomorphology from such schematic diagrams alone; if they are given topographical maps it is like giving a child the ledgers of a large business for him to learn arithmetic; the comparison of this schematic material with reality, by means of maps and pictures as well as by fieldwork, can come later. To this end he published a special atlas for teaching geomorphology, the German edition of which was produced by Östreich.* In some measure this is true; greater simplicity facilitates understanding. Simplified drawings have always been used, even if not as extensively and in so perfected a fashion. But to make an intensive study of drawings before one has made a direct study of nature easily lends to the latter being bypassed altogether, especially when the drawing is not merely a simplification, an abstraction of reality, but contains genetical concepts. Davis's drawings are the product of his conceptual system and portray the surface of the land according to the way he thought it originated. Therefore the same is true of these diagrams as of his method of presentation: they stand or fall on whether his deductive theoretical concepts are valid or not. The adherents of Davisian orthodoxy believe in them, the rest of us remain sceptical of many of them and consider them pedagogically questionable. Because of its positiveness, a drawing impresses the senses far more and fixes itself more firmly in the memory than does the spoken word. But precisely because it does so, a drawing is especially hazardous when it is wrong.

Geological Table for South Germany

(to show the stratigraphical position of strata referred to in the text)

ERA	PERIOD	ROCK SERIES	COMPOSITION AND FORM
Quaternary	Holocene		alluvium (river deposits)
Quaternary	Pleistocene		porous loess and diluvium (sand and gravels), impervious moraines and clays
Tertiary	Pliocene Miocene Oligocene Eocene	Molasse Flysch	conglomerates, lignites, impervious clays, porous sands, resistant basaltic eruptives
Secondary	Cretaceous	Quadersandstein	cliff- and plateau-forming quartz sandstone, with clays and conglomerates
Secondary	Jurassic	White Jura (Malm) Brown Jura (Dogger) Black Jura (Lias)	scarp-forming, porous limestone and dolomite, with shales, marls, slates and clays
Secondary	Triassic	Keuper Muschelkalk-Wellenkalk Bunter Sandstone	scarp-forming sandstone and limestone, with mainly pervious shales
Primary	Permian	Rotliegendes	impervious conglomerates, shales and clays
Primary	Archaen and Palaeozoic		impervious gneiss, granite and schist of the basement complex.

Author's and Translator's Notes and Bibliographic References

The following contractions are used below in both the author's and translator's notes and bibliographic references.

A.F.N.F.	Abderhalden's Fortschritte der Naturwissenschaftliche Forschung.
A.G.	Annales de Géographie (Paris).
A.P.C.	Annalen der Physik und Chemie (Berlin).
A.R.G.S.	Annual Report, United States Geological Survey.
A.S.A.W. (m-p)	Abhandlungen der Kgl. Sächsischen Gesellschaft (Akademie) der Wissenschaften, (Math-Phys Klasse) (Leipzig).
A.S.P.N.	Archives des sciences physiques et naturelles (Geneva).
B.G.I.	Bulletin of the Geological Institute (Uppsala).
B.M.C.Z.	Bulletin of the Museum of Comparative Zoology (Harvard).
B.O.L.	Beiträge zur Oberrheinische Landeskunde.
C.I.G.	Congrès International de Géographie.
C.I.G.E.	Congrès International de Géologie.
E.L.A.	Erdgeschichtliche und landeskundliche Abhandlungen aus Schwaben und Franken (Tübingen).
F.G.P.	Fortschritte der Geologie und Paläontologie (Berlin).
F.L.V.	Forschungen zur Deutschen Landes- und Volkskunde (Stuttgart–Bad Godesberg).
G.A.	Geographische Abhandlungen (Leipzig).
G.AN.	Geografiska Annaler (Stockholm).
G.ANZ.	Geographische Anzeiger (Gotha).
G.B.	Geographische Bausteine (Gotha).
G.J.	Geographical Journal (London).
G.JB.	Geographisches Jahrbuch (Gotha).
G.M.	Geological Magazine (London).
G.R.	Geologische Rundschau (Leipzig).
G.REV.	Geographical Review (New York).
G.Z.	Geographische Zeitschrift (Leipzig).
J.G.R.A.	Jahrbuch der geologischen Reiches-Anstalt (Vienna).
J.O.G.V.	Jahresberichte und Mitteilungen des Oberrheinischen Geologischen Vereins (Stuttgart).
J.S.A.K.	Jahrbuch der Schweizer Alpenklub.
K.Z.	Kartographische und Schulgeographische Zeitschrift (Vienna).

M.G.H.	Mitteilungen der Geographischen Gesellschaft (Hamburg).
M.G.S.	Memoir of the Geological Survey of Great Britain.
M.G.S.I.	Memoir of the Geological Survey of India.
M.V.E.L.	Mitteilungen der Verein für Erdkunde (Leipzig).
N.G.M.	National Geographic Magazine (Washington D.C.).
N.N.G.Z.	Neujahresblatt der Naturforschende Gesellschaft (Zürich).
P.A.A.S.	Proceedings of the American Association for the Advancement of Science (Washington D.C.).
P.A.S.B.	Proceedings of the American Academy of Arts and Science (Boston).
P.G.A.	Proceedings of the Geologists Association (London).
P.G.S.	Proceedings of the Geological Society (London).
P.M.L.	Philosophical Magazine (London).
P.M. (Erg-h)	Petermanns Geographische Mitteilungen (Ergänzungsheft) (Gotha).
Q.J.G.S.	Quarterly Journal of the Geological Society (London).
R.B.A.	Report of the British Association for the Advancement of Science (London).
R.G.A.	Revue de Géographie Annuelle (Paris).
S.P.A.W. (m-p)	Sitzungsberichte der Kgl. Preuss. (Akademie) der Wissenschaften, (Math-Phys. Klasse) (Berlin).
S.S.A.W. (m-p)	Sitzungsberichte der Kgl. Sächsischen Gesellschaft (Akademie) der Wissenschaften, (Math-Phys. Klasse) (Leipzig).
S.V.N.K.	Schriften der Verein zur Verbreitung Naturwissenschaftliche Kenntnisse (Vienna).
S.W.A.K. (m-n)	Sitzungsberichte der Osterreichische Akademie der Wissenschaften, (Math-Naturwiss. Klasse) (Vienna).
T.C.S.	Transactions of the Cavendish Society (London).
V.D.G.	Verhandlungen der Deutschen Geographentages.
V.G.D.N.	Verhandlungen der Gesellschaft Deutschen Naturforscher und Arzte.
V.G.E.	Verhandlungen der Gesellschaft für Erdkunde (Berlin).
V.N.G.	Vierteljahrschrift der Naturforschende Gesellschaft (Zürich).
Z.D.G.G.	Zeitschrift der Deutschen Geologischen Gesellschaft (Berlin).
Z.G.	Zeitschrift für Geomorphologie (Berlin).
Z.GL.	Zeitschrift für Gletscherkunde (Berlin).
Z.G.E.	Zeitschrift der Gesellschaft für Erdkunde (Berlin).
Z.M.G.P.	Zentralblatt für Mineralogie, Geologie und Paläontologie (Stuttgart).
Z.W.G.	Zeitschrift für Wissenschaftliche Geographie (Lahr/Weimar Vienna).

Author's Notes and Bibliographic References

(1) Walther Penck, *Morphological Analysis of Land Forms, a contribution to Physical Geology* (London 1953). Earlier in the article: *Wesen und Grundlagen der morphologischen Analyse* (S.S.A.W. 1920, LXXII).
(2) The state of geomorphological knowledge was summarised by G. de la Noë and E. de Margerie in their *Les formes du terrain* (Paris 1888), and in A. Penck, *Morphologie der Erdoberflache* (2 vols., Stuttgart 1894), (a treatment showing great command of the material in a way that considers geomorphology the systematic study of a particular kind of phenomena), and more briefly in A. Supan, *Grundzüge der physischen Erdkunde* (Leipzig 1884).
(3) First in a series of separate articles, of which those containing material of more general importance were collected together in his *Geographical Essays* (Boston 1909), and later in a comprehensive work in German *Die Erklärende Beschreibung der Landformen* (Leipzig 1912), as well as in a smaller textbook *Physical Geography* (Boston 1898) and its German edition by G. Braun, *Grundzüge der Physiogeographie* (Berlin 1911 and 1915). Rühl's articles: *Eine neue Methode der Morphologie* (A.F.N.F. 1912, VI), and a more recent article by Davis himself: *The explanatory description of landforms*, in the *Festschrift für J. Cvijić* (Belgrade 1924, pp. 287 ff.), give a good review of his teaching.
(4) In various lectures and in his *Leçons de Géographie Physique* (Paris 1896 and 1909).
(5) Even the fourth edition of his work: *Traité de géographie physique*, which appeared in 1927, in still entirely within the framework of Davis's teaching.
(6) My critique began, apart from more occasional earlier statements, with an article on: *Die Terminologie der Oberflächenformen* (G.Z. 1911, 17), and was continued in succeeding years in a series of articles in the *Geographische Zeitschrift*. A similar critique pervades S. Passarge's *Physiologische Morphologie* (M.G.H. 1912, 26). In the last edition of his *Grundzüge der physischen Erdkunde* (Leipzig 1916), Supan also discarded Davis's approach.
(7) The progress of geomorphology was considered by J. Sölch in his *Die Fortschritte unserer Kenntnis der exogenen Kräfe 1914–1924* (G. JB. 1924/25, 40, pp. 100–272). Recent comprehensive presentations of geomorphology have been given by S. Passarge, *Grundlagen der Landschaftskunde* (vol. III, Hamburg 1920), and A. Philippson, *Grundzüge der allgemeinen Geographie* (vol. II, Leipzig 1924).
(8) Yet Davis declared in a review of the first edition of the present work,

that he disregarded them for didactic reasons alone, since there was no opportunity to show minor landforms in the field during the winter lectures in Berlin, from which his book *Die Erklärende Beschreibung der Landformen*, derived. But the book claims to be a comprehensive treatment; and there is scarcely any reference to minor landforms in the book he wrote in collaboration with G. Braun *Grundzüge der Physiogeographie* (Berlin 1911) either.

(9) Passarge provides an exhaustive consideration of these processes in the third volume of his *Grundlagen der Landschaftskunde* (Leipzig 1920). But he separates sheet wash (*abspülung*) too sharply from mass-movement (*Bodenversetzungen*), and does not take enough account of rinsing.

(10) 'Dell' is no newly devised term, as is sometimes thought, but an old one found on many maps. I used it in my book on Saxon Switzerland, and it is to be found in Grimm's Dictionary. C. Heim described the phenomenon as early as 1791. W. Penck's *Korrasionstalungen*, which he saw as being formed by the movement of dry débris, are roughly the same thing.

(11) In summary fashion in the *Düsseldorfer geographischen Vorträgen*, part III (Breslau 1927, p. 70).

(12) *Die Denudation in der Wüste und ihre geologische Bedeutung* (A.S.A.W. (m) 1891, 16), and later *Das Gesetz der Wüstenbildung* (4th Edit., Leipzig 1924).

(13) The *Düsseldorfer geographischen Vorträge*, part III: *Morphologie der Klimazonen* (Breslau 1927), gives a review of the present state of our knowledge.

(14) In *Wesen und Grundlagen der morphologischen Analyse* (S.S.A.W. 1920, LXXII, p. 66).

(15) *See* my *Gebirgsbau und Oberflächengestaltung der Sächsischen Schweiz* (Stuttgart 1887, p. 62).

(16) *Die Kordillere von Bogotá* (P.M. (Erg-h) 1892, 104, p. 45 ff.).

(17) W. Penck has read rather carelessly when he asserts (*Morphological Analysis of Landforms*, London 1953, p. 177) that my benchlands, and ledgelike denudation terraces of valleys, are identical.

(18) P.A.A.S. 1884, XXXIII.

(19) G.J. 1899, XIV, p. 481. *Geographical Essays* (Boston 1909, p. 249).

(20) My criticism is partly that of Passarge (*Physiologische Morphologie*, M.G.H. 1912, 26, pp. 17 and 149 f.) but began independently of his; I expressed it in brief comments a long time ago.

(21) When Davis upbraids me and other critics for wrongly putting the word 'time' into his mouth, he forgets that he formerly (for example in *Geographical Essays*, Boston 1909, p. 249) always spoke of time, and has only recently withdrawn this expression. In his book with Braun *Grundzüge der Physiogeographie* (2nd Edit., Berlin 1915/17, vol. II, p. 5), stage is still equated with time elapsed.

(22) Z.G.E. 1912, XLVII, p. 298.

(23) That this too is about to become even, as is asserted in W. M. Davis,

G. Braun, *Grundzüge der Physiogeographie* (2nd Edit., Berlin 1915/17, vol. II, p. 102), is incorrect.

(24) Even Davis himself (*Die Erklärende Beschreibung der Landformen*, Leipzig 1912, p. 183) and Rühl (*Eine neue Methode der Morphologie*, A.F.N.F., 1912, V, I, p. 89) later acknowledged that the age of valley bottom, that of valley sides, and the development of the drainage pattern are three different things. It follows logically from this that we should stop designating valleys by age, and give up the terminology based on age, or restrict it to the characterisation of valley bottoms. But valleys are still described in their entirety as young, mature or old, without any regard for differences in the way they have developed.

(25) We have no justification for using the American form 'canyon' in German.

(26) The use of such expressions, originally restricted to one locality or to a particular natural or linguistic region anyhow—such as in this case the Spanish-American Cordillera—as terms, is not of course a question of correctness but of expediency. It therefore depends upon scientific tact. In this instance tact has not been used. W. von Lozinski's statement (J.G.R.A. 1909, LIX, p. 6 and 39 ff.) is of no significance in the study of canyons, since it lacks an exact definition of terms. For the same reason their description in Supan's *Grundzüge der physischen Erdkunde* (5th Edit., Leipzig, p. 524) goes awry. The unfortunate tendency to describe any valley incised into a plateau, and without a valley bottom, as a canyon, also prevails in Davis's school. Passarge considers even the valley of wooded mountains canyons.

(27) Since this gives canyons a certain resemblance to a narrow 'U', they have lately been called U-shaped valleys. Yet again this introduces unnecessary error into scientific terminology; the description U-shaped has long been given to valleys of quite another form and origin.

(28) W. von Lozinski and A. Supan contend that in his description of the Hawaiian Islands (A.R.G.S., 1882/83, IV, p. 216 f.) Dutton withdrew his original explanation of canyons as the product of a dry climate. That is not true; he maintains much more explicitly that it was the correct explanation, and adds only that in moist climate permeable rock produces similar landforms to those of a dry climate. Thus he gives the same explanation as that to which I and others were led somewhat later through studying the valleys of the Quadersandstein Mountains—a fact I earlier overlooked.

(29) *Physiologische Morphologie* (M.G.H., 1912, 26, p. 150). But Richthofen used this expression in quite another sense, *Führer für Forschungsreisende* (Hanover 1886, p. 146).

(30) The term 'consequent' introduced by Powell does not correspond with the true meaning of this word, and is therefore best avoided.

(31) 'Subsequent valleys' (*Nachfolgender Täler*) in contradistinction to 'succeeding valleys' (*Folgetälern*) after Penck, is not to be recommended either, since in colloquial language 'to succeed' and 'to follow' have the same meaning.

(32) Davis restricted the expression 'subsequent' to a particular kind of valley developed later. In doing so he relies on J. Beete Jukes (Q.J.G.S. 1862, 18, p. 400); but in the relevant place Beete Jukes says that certain longitudinal valleys are of subsequent origin; that is certainly a different thing. Davis then has to find other expressions such as 'obsequent', etc., for the transverse valleys which develop later. Such a combination of the direction of valleys and the time at which they originate in a single expression is inexpedient. In their translation into German, O. Krummel and H. Fischer retained only valley direction.

(33) V.G.E. 1889, 3.

(34) *See* the statement by H. Schmitthenner (G.Z. 1916, 22, pp. 259 ff.).

(35) De Martonne's choice of the term *plateforme structurale* to contrast it with the *plateformes d'erosion* seems less clear to me; benchlands, unlike the *plateformes structurale*, are not the direct result of structure but a feature of denudation (not erosion in the strict sense) in which structure also plays a role.

(36) R. Gradmann, *Das Schichtstufenland* (Z.G.E. 1919, 54, pp. 133 ff.), notes that Gümbell had already observed this.

(37) G. Z. (1903, 9, pp. 621 ff.), where I have also reviewed the theory of scarplands.

(38) Evidence that they depend upon rock types has quite recently been presented in a painstaking study by Lamprecht.

(39) R. Gradmann has adequately refuted this interpretation in the paper referred to.

(40) *Die Entstehung der Stufenlandscaft* (G.Z. 1920, 26, pp. 207 ff.). Philippson's objections (*Grundzüge der allgemeinen Geographie*, Leipzig 1924, vol. II, pp. 342 ff.) seem invalid to me.

(41) Philippson disputes this, or at least only concedes it to a lesser degree, but without giving his reasons.

(42) *Drainage of Cuestas* (P.G.A. 1899, XVI).

(43) *Das Schichtstufenland* (Z.G.E. 1919, 54, pp. 113 ff.).

(44) S.W.A.K., (m-n) (1896, CV, pp. 147 ff., and G.Z. 1897, 3, p. 47).

(45) *Physiologische Morphologie* (M.G.H. 1912, 26, p. 42).

(46) The term should be restricted to such hills, for which it was created, and not extended to other isolated hills such as residuals.

(47) I am indebted to my friend Schmitthenner for information which has greatly clarified this problem.

(48) R.B.A. (1847, p. 66) and M.G.S. (vol. 1, p. 297), later repeated in *Physical Geology and Geography of Great Britain* (6th Edit., London 1894, p. 346 f.).

(49) F. von Richthofen, *China* (vol. 11, Berlin 1884, p. 766 f.); *idem., Führer für Forschungsreisende* (Hanover 1886, pp. 353 ff.).

(50) *Erdgeschichte* (vol. 1, 1st Edit., Leipzig/Vienna 1886, pp. 447 and 484 f.).

(51) S.W.A.K. (1886/87, 27, pp. 431 ff.) and in a lecture *Das Endziel der Erosion und Denudation*, at the Berlin Geographical Congress 1889 (V.D.G. 8, Berlin 1889).

(52) N.G.M. (1889, 1, pp. 183 ff.), reprinted in *Geographical Essays* (Boston 1909, pp. 413 ff.).
(53) Richthofen also later accepted this sub-aerial formation as well as marine, as possible. (M. Neumayr, *Anleitung zu wissenschaflichen Beobachtungen auf Reisen* (2nd Edit., Berlin 1888, vol, 1, p. 247).)
(54) The term peneplain has a wider meaning colloquially, and embraces benchlands as well. The words 'land surface', as used by Philippi and others, does not imply planation and need not be understood as a planed surface. Philippson's term 'denudation surface' is also too wide in meaning; for such a surface need be neither a plain nor a peneplain.
(55) J.O.G.V. (1919, VIII, p. 123).
(56) Reck, in the Z.D.G.G. (1912, p. 83).
(57) Z.D.G.G. (1910, LXII, pp. 305 ff.).
(58) G. Braun: *Deutschland* (Berlin 1916, p. 18).
(59) A. G. (1908, XVII, pp. 205 ff.).
(60) A crossing of the Appeninnes from Modena to Lucca in the summer of 1912 showed us a true cuesta landscape rather than a peneplain as Braun has claimed. Brückner's re-uplifted peneplain in the Swiss Jura was no more evident to us in 1911 (cf. G.Z. 1912, 18, pp. 515 ff.); the latter was refuted at the same time by Martin (R.G.A. 1911, p. 219).
(61) Machatschek carried out such a study in the mountains of the Alpine system (Z.G.E. 1916, LI, pp. 602 f., and 675 f.); although he is critical of many assertions, his criticism does not seem to me to go far enough.
(62) *See* the exhaustive discussion in A. Penck, *Morphologie der Erdoberfläche* (Stuttgart 1894, vol. II, pp. 616 ff.) and more recently in his article *Die Gipfelflur der Alpen* (S.P.A.W. (m-p) 1919, XVI, p. 256).
(63) I cannot understand how Philippson can say that I oppose the reality of peneplains solely on the basis of the long period needed for their formation (*Grundzüge der allgemeinen Geographie*, Leipzig 1923/1924, vol. II, p. 530) in view of the preceding exhaustive citation of the evidence which appeared even in the 1st Edition of the present work.
(64) G.Z. (1913, 19, pp. 185 ff.).
(65) *See* Eduard Richter's *Geomorphologische Untersuchungen in den Hochalpen* (P.M. (Erg-h) 1900, 132) and my *Gebirgsbau und Oberflachengestaltung der Sächsischen Schweiz* (Stuttgart 1887, especially p. 96).
(66) *Eine neue Methode der Morphologie* (A.F.N.F. 1912, VI, p. 80).
(67) *op. cit.* (p. 99).
(68) *op. cit.* (p. 110).
(69) W. Penck and others argue against this use of the term 'internal build' in the wider sense. Of course the question is only of terminological importance; but terminology aids or hinders knowledge. We need a word for the overall result of internal (endogenic) processes. The architecture of a house consists not only of the position of individual bricks, but also of their composition, and the plan and style of the whole. The same is true of the earth's land surface. Widespread uplift (epeirogenic move-

ment) also changes the position of strata, except that we can only infer this, not observe it directly. The word 'structure' is best left with its narrower meaning.

(70) *Die Erklärende Beschreibung der Landformen* (Leipzig 1912, p. 305).

(71) The terms homogeneous and heterogeneous seem to me to meet the nature of the case better than those proposed by Passarge (*Physiologische Morphologie*, M.G.H. 1912, 26, p. 119), 'harmonious' and 'disharmonious', which are not causal but teleological. They refer to the aesthetic result, not the origin. The term later proposed by Salomon, 'dead landscapes', is too easily misunderstood; a man is not dead when he is subjected to different formative influences. Such landscapes could perhaps be named 'fossil', but the expression has long since been given up in its aesthetic sense.

(72) Dacqué gives a good critical review in his *Grundlagen der Paläogeographie* (Jena 1915).

(73) Z.G.E. (1911, XLVI, pp. 479 ff.).

(74) Passarge also considers this problem (*Physiologische Morphologie*, M.G.H. 1912, 26, Chapter V) as well as A. Penck in *Die Formen der Landoberfläche und Verschiebungen der Klimagürtel* (S.P.A.W. (m-p) 1913, XLVI, pp. 577 ff.). Part three of the *Düsseldorfer geographischen Vorträge* (Breslau 1927) is devoted to the geomorphology of climate regions.

(75) This idea remains even after the exhaustive exposition by W. Soergel, *Löss, Eiszeiten und Paläolithische Kultur* (Jena 1919).

(76) S.P.A.W. (m-p) (1900, XXXIV, p. 246).

(77) *Grundlagen der Landschaftskunde* (vol. III, Hamburg 1920, p. 100).

(78) See A. Penck, *Klimaklassifikation auf physiogeographischer Grundlage* (S.P.A.W. (m-p), 1910, XLIII, pp. 236 ff.). See my review in the G.Z. (1910, 16, p. 645).

(79) *Führer für Forschungsreisende* (Hanover 1886, pp. 304 ff.).

(80) Z.W.G. (1885, V, pp. 245 ff.).

(81) *Geology of the U.S. Exploring (Wilkes) Expedition* (Philadelphia 1849).

(82) *Richthofen-Festschrift* (Berlin 1893, pp. 1 ff.).

(83) D. W. Johnson, *Shore processes and shoreline development* (New York 1919).

(84) *Eine neue Methode der Morphologie*, A.F.N.F. (1912, VI, pp. 67 ff. and p. 81).

(85) *Grundlagen der Landschaftskunde*, vol. 1 (Hamburg 1919, pp. 23 ff.).

(86) In view of this I cannot understand how Walther Penck can assert that I do not grant deduction any validity at all; I am only opposed to its excessive use.

(87) *Eine neue Methode der Morphologie*, A.F.N.F. (1912, VI, p. 129 ff.).

(88) Davis has himself provided such an analysis, even to a laborious extent, in his study of the Welsh Uplands.

(89) *Die Erklärende Beschreibung der Landformen* (Leipzig 1912, Foreword p. vii f.).
(90) *See* my book *Die Geographie, ihr Wesen, ihre Geschichte und ihre Methoden* (Leipzig 1927, pp. 388 ff.).
(91) *See* F. Machatschek, in K.Z. (1917, 3/4, pp. 1 ff.), N. Creutzburg, in P.M. (1925, 71, p. 70), H. Weber, in P.M. (1924, 70, p. 71 f.), and R. Mayer, in Z.G. (1926, 32, pp. 160 ff.).
(92) *See* for example A.R.G.S. vol. IX (Washington 1875, plates 36 and 42).

Translator's Notes and Bibliographic References

Author's Foreword, page xxi

(i) The lengthy phrase *'Lehre von den Oberflächenformen des Landes'* is translated here simply as 'geomorphology', although we must then use this same word for 'Morphologie' when it occurs later on the same page. Hettner is at pains to distinguish between a systematic science of landforms as such, geomorphology (p. 3), and the geographical study of the earth's surface features. It is noteworthy that in this second edition of his work Hettner discarded the book's original subtitle referring to the surface features of the lands, 'their investigation and portrayal', in favour of an explicit reference to geomorphology. I have therefore tried to reserve the word 'geomorphology' to translate *Morphologie* alone, but this has not always been possible.

page xxii

(ii) Hettner is here referring to W. M. Davis's review of the first edition of his work in the *Geographical Review* (1923, vol. XIII, p. 318–21), under the surprising title 'The Explanatory Description of Landforms'.
(iii) A. Hettner, *Die Geographie, ihr Wesen, ihre Geschichte und ihre Methoden* (Leipzig, 1927).

Introduction, page xxiv

(1) Although it is convenient to render *Bauplan, Baustil* (Ch. VII, p. 85) and *Gebirgsbau* (44, 53) as 'structural plan', 'structural style' and 'mountain structure' respectively, I have consistently translated the expression *innere Bau* in a quite literal way as 'internal build'. It is an essential feature of Hettner's opposition to W. M. Davis, as he argues it on pages 95–7 and 141–2, that the geographer, in his appraisal of the surface on which exogenic processes work, must take account of more than the way rocks are disposed, their dip, strike, bedding and the like (those characteristics Davis grouped under the term *feineren struktur* in his *Die Erklärende Beschreibung der Landformen* (Leipzig 1912, p. 65), and Hettner, to avoid using the word *struktur* except in a purely petrological sense (p. 9), used the term *lagerungsverhältnisse* (p. 8); he must, in Hettner's view, also try to reconstruct what Hettner calls the tectonic surface, the product of the crust's deformative history of warping, uplift, subsidence and vulcanism. The *innere Bau* of an area is, for Hettner, the outcome of both the nature and arrangement of the rocks of which the area is built, and of the constructional history it has under-

Translator's Notes and Bibliographic References

gone. In arguing that by *struktur* Davis means only the way rocks are disposed (p. 95) Hettner is less than just. Although Davis certainly excluded crustal movement from his concept of structure, he did include in it the purely formal construction (one akin to Hettner's tectonic surface) he called the 'original surface' (*op. cit.*, p. 143) and seems to have distinguished this from what he terms the *inneren struktur* (*op. cit.*, p. xv). But Hettner is certainly justified in maintaining his position (p. 95 and note 70) against W. Penck's use of the two terms 'structure' and 'internal build' synonymously in the narrower sense to mean only rock disposition (*Morphological Analysis of Land Forms*, London 1953, p. 19 and note 21). It is of interest to note that contrary to Penck's view, a modern geologist, L. U. de Sitter, treats of both the static disposition and the dynamic deformation of rocks in his *Structural Geology* (New York 1956).

page xxv

(2) The principle of actualism. Hettner later (p. 130) refers to workers other than Lyell who played perhaps an even more important role than he in the introduction of both the actualist and gradualist viewpoints into geomorphology. See translator's note 136.
(3) M. Neumayr, see author's note 50. A third edition of Neumayr's two-volume work was published in 1921. J. Walther: see author's note 12. The first edition of Walther's book *Das Gesetz der Wüstenbildung* was published in 1900.
(4) The text of W. Penck's *Die Morphologische Analyse* (Stuttgart 1924) referred to is the translation by H. Czech and K. C. Boswell entitled *Morphological Analysis of Land Forms* (London 1953).

page xxvii

(5) A correct translation of *Kleinformen* and *Grossformen* here is of some importance. We must later (pp. 138-40) go on to translate the terms *Hauptformen* and *Gesamtformen* so as to set up a hierarchy of categories into which landforms may be grouped according to both their relative size and the degree to which they are independent of each other. The point of Chapter I is lost unless we appreciate that by 'minor landforms' Hettner means features such as a boulder-field, an earth-pyramid, or a cavern, in other words landforms of the seventh order according to the classification of A. Cailleux and J. Tricart in *Le problème de la classification des faits Géomorphologiques* (A.G. 1956 LXV, pp. 162-186). W. M. Davis, on the other hand, would appear to describe all landforms smaller than those of the first or second order as *kleineren*—*Die Erklärende Beschreibung der Landformen* (Leipzig 1912, p. 23). It is because of the importance Hettner attaches to the study of what might be termed micro-relief that I have generally translated *formen* as 'features' not 'landforms'.

G*

178   *The Surface Features of the Land*

page xxviii

(6) O. Peschel, *Neue Probleme der Vergleichenden Erdkunde* (1st Ed., Leipzig 1869).

(7) This is apparently a reference to Davis's review of the 1st edition of Hettner's work. *See* translator's note (ii).

(8) Hettner invariably qualifies the expression 'surface of the earth' (*Erdoberfläche*) and 'earth crust' (*Erdrinde*) with the word *festen*; the title of the book itself contains the word *Festlandes*. As he explains in the preface to his *Vergleichende Länderkunde* (Leipzig/Berlin 1933, p. iii), Hettner avoids the word 'continents' when speaking of the earth's land surface; he wishes to include islands and smaller fragments of the unsubmerged crust of the earth in his treatment. It may be presumed that W. M. Davis had a similar intention when he entitled his paper republished as Chapter 3 in *Geographical Essays* (Boston 1909), 'The Physical Geography of the Lands'.

page xxix

(9) L. Rütimeyer, *Uber Tal und Seebildung* (Basle 1869).

(10) Hettner rarely uses the word 'denudation' alone, and then only when it is clear from the context (as on pp. 19, 28 and 49 for example) that he is referring to the way in which the removal of material by mass-movement and sheet wash lays bare a slope to fresh weathering and disintegration. At other times, as here, he speaks of degradation (*abtragung*) by surface or sheet denudation in contrast with that brought about by linear erosion. In this way, Hettner avoids the confusion which can arise when denudation is used in both the narrower sense, as it is in German generally, and in a broader sense to include weathering, erosion both sheet and linear, and deposition—for example in L. B. Leopold, M. G. Wolman, J. P. Miller, *Fluvial Processes in Geomorphology* (San Francisco 1964, Ch. III). By translating both *abtragung* and *denudation* as simply 'denudation', H. Czech and K. C. Boswell use the word in the narrower and broader sense indiscriminately, and so obscure W. Penck's thought in an important respect (*Morphological Analysis of Land Forms*, London 1953, in particular Ch. IV). It should be carefully noted that Hettner's distinction between 'denudation' and 'degradation' is not exactly that of W. M. Davis (*Base-Level, Grade and Peneplain*, J.G. 1902, X, p. 107).

(11) F. von Richthofen, *see* author's note 29.

page xxxi

(12) I know of no adequate term in English to translate *Rumpffläche*—the extensive planed-off and almost level surface of a worn-down remnant upland. I have therefore rendered it simply as 'remnant surface'. It cannot be translated as 'peneplain' (*Fastebene*), for certainly not all, if any, such remnant surfaces are peneplains in the Davisian sense (Ch. VI). Though well established, the expression 'erosion surface' can as well be

Translator's Notes and Bibliographic References 179

applied to a steep valley side as a nearly flat platform, and it does not suggest that the feature is the surface of an old remnant mass, the *Rumpf*. The descriptive term 'upland plain' (S. W. Wooldridge, *The Geographer as Scientist*, London 1956, p. 129) is inappropriate since the German word conveys nothing as to elevation, although admittedly the features in question are usually at some considerable height above sea level; but in any case I have reserved 'upland plain' for Hettner's equally descriptive *Hochfläche* (p. 18). However we translate *Rumpffläche* our word must subsume the terms we shall need to use to distinguish a level truncating the strata across which it has been formed, one that conforms with the top of a resistant master-stratum, a peneplain, a strand-flat, the outcome of lateral river corrosion, arid planation and advanced corrie amalgamation (*see* Ch. VI).

Chapter 1, page 1

(13) *See* geological table, p. 166. In England, France and Belgium the Upper Cretaceous is composed largely of soft chalk. But eastwards and southwards into central Europe it is of increasingly sandier rocks, until in the mountains through which the Elbe River cuts in a magnificent gorge, the area generally known as Saxon Switzerland (*Sächsischen Schweiz*), it is represented by a thick quarzitic sandstone, the Quadersandstein, so-called from its tendency to break up into joint-bounded blocks or '*quader*'. (*See Absonderungsformen*, p. 8 of the present work.) This uniformly porous, massive and jointed sandstone gives the region a characteristic scenery of vertical rock faces rising above bench-like levels and deep chasms. It was in this area that Hettner carried out his first geomorphological researches, work of considerable importance for his later thought. (*See* p. 63 and his note (15).)

page 2

(14) The word *Boden* normally has a variety of meanings: ground, bottom, floor, soil. When it is used alone (p. 4) I have translated it as 'soil'. But when it is used in conjunction with other words, such as in *Talboden* (p. 25) or as here, *Bodenbeschaffenheit*, I have tendered it as 'floor' or used a phrase which avoids an actual translation of the word.

page 3

(15) E. Chaix du Bois, *Le Pont des Oulles* (La Géographie 1903, VIII) and *idem La topographie du désert de Platé* (Le Globe, Geneva 1895).

page 4

(16) Presumably a reference to K. E. A. von Hoff's *Geschichte der durch die Uberlieferung nachgewiesenen Veränderung der Erdoberfläche* (3 vols., Gotha 1825–41), H. Müller's *Die Befruchtung der blumen durch Insekten und die gegenseitigen anpassungen beides* (Leipzig, 1873), and J. Lubbock's *Ants, bees and wasps* (1st Ed., London 1886). Examples of

detailed observation by the authors Hettner names are: S. Passarge, *Verwitterung und Abtragung in den Steppen und Wüsten Algeriens* (G.Z. 1909, XV), K. Sapper, *Uber Abtragungsvorgänge in den regenfeuchten Tropen und ihre morphologischen Wirkungen* (G.Z. 1914, XX), J. Walther, *Windkanter aus der Libyschen Wüste* (Z.D.G.G. 1911, LXIII).

page 7

(17) G. Götzinger, *Beiträge zur Entstehung der Bergrückenformen* (G.A. 1907, IX).
(18) H. Schmitthenner, *Die Entstehung der Dellen und ihre morphologische Bedeutung* (Z.G. 1925/26, 1). *See* also Schmitthenner's paper to which Hettner refers in his note 40.
(19) See translator's note 12.

page 8

(20) B. Cotta, *Deutschland's Boden* (2nd Ed., Leipzig 1858).
(21) *See* translator's note 1, and p. 89.

page 9

(22) It is impossible to know whether by *gleichen* Hettner means 'identical' or merely 'similar'; but he later unequivocally speaks of climates being similar (*ähnliche*). My rendering here will allow for the possibility that Hettner is thinking of those climates which, though quite different in origin, have virtually identical characteristics and virtually the same effect on the land surface. *See*, for example, p. 144. The same considerations would apply to *gleiche Eigenschaften* and *gleichen Vorgänge* in the next paragraph.

page 10

(23) G. Schweinfurth, *Die Umgebung von Heluan als Beispiel der Wüstendenudation* (1895/96). See also the bibliography appended to J. Walther, *Das Gesetz der Wüstenbildung* (1st Ed., Leipzig 1900).

page 11

(24) *See* translator's note 10.
(25) Hettner's use of the word *Eiszeit* is most careless. Following the expression introduced by L. Agassiz, I have translated *Eiszeit* by 'Glacial Epoch', but *Die Eiszeiten* (p. 110) by 'glacial periods'. In some contexts, for example on p. 119, it is clear that by *der Eiszeit* and the *letztere Eiszeit* Hettner is referring only to the Last Glaciation, and by *Haupteiszeit* (p. 39) to the local glacial maximum of an area and not to any particular glacial period. But on p. 110 he uses *Eiszeit* to mean both the Glacial Epoch as a whole and glacial as distinct from interglacial periods (*Zwischeneiszeiten*).

page 12

(26) E. Obst, *Die Oberflächengestaltung der sächsisch-böhemischen Kreideablagerungen* (M.G.H. 1909, XXIII).

Chapter II, page 14

(27) The extent to which C. Lyell departed from the views of Hutton may be judged by reference, for example, to Lyell's *Principles of Geology* (1st Ed., London 1833, vol. III, pp. 294–320). *See* also: J. Hutton, *Theory of the Earth* (2 vols., Edinburgh 1795), J. Playfair, *Illustrations of the Huttonian Theory of the Earth* (Edinburgh 1802), G. Greenwood, *Rain and Rivers, or Hutton and Playfair against Lyell and all comers* (1st Ed., London 1857), J. B. Jukes, *On the mode of formation of some of the river valleys in the South of Ireland* (Q.J.G.S. 1862, 18, pp. 378–403), a paper to which Hettner refers in his note 32, O. Peschel, *Die Thalbildungen*, in his *Neue Probleme der Vergleichenden Erdkunde* (3rd Ed., Leipzig 1878, pp. 150–164), L. Rütimeyer, *see* translator's note 9.

page 15

(28) W. Deecke, *Der Zusammenhang von Flusslauf und Tektonik, dargestellt an den Flussen Süddeutschlands* (F.G.P. 1926, 16).

page 18

(29) G. K. Gilbert, *Report on the Geology of the Henry Mts* (U.S. Geogr. and Geol. Survey, Washington D.C. 1877, p. 160). F. von Richthofen, *Führer für Forschungsreisende* (Hanover 1901, Ch. 6, pp. 130–204). G. de la Nöe, E. de Margerie, *see* author's note 2. A. Penck, *Das Endziel der Erosion und Denudation* (V.D.G. 8, Berlin 1889). *See* author's note 51 and Hettner's reference to this work on p. 23. A. Philippson, *Die Erosion des fliessenden Wassers und ihre Einfluss auf die Landschaftsformen* (G.B. 1914, 7), and his paper *Ein Beitrag zur Erosionstheorie* (P.M. 1886, XXXI, pp. 67–79). A. Hettner, *Die Arbeit des fliessenden Wasser* (G.Z. 1910, XVI, pp. 365–384).

page 19

(30) J. Brunhes, *Le travail des eaux courantes: la tactique des tourbillons* (A.G. 1904, 13).

page 20

(31) *See* translator's note 14. On page 25 Hettner explains why he distinguishes between the 'floor' (*Talboden*) and the 'bottom' (*Talsohle*) of a valley.

page 22

(32) A Surell, *Étude sur les torrents des Hautes Alpes* (Paris 1841). H. Baulig, *La notion de profil d'équilibre histoire et critique* (C.I. Cairo 1925, 3, pp. 51–63). J. W. Powell, *Exploration of the Colorado River of*

the West and its tributaries (Washington D.C. 1875). G. K. Gilbert, *see* translator's note 29. A. Heim, *Uber die Erosion im Gebiete der Reuss* (J.S.A.K. 1878–79). A. Philippson, *see* translator's note 29.

page 23

(33) A. Penck, *see* author's note 51 and translator's note 29.

page 24

(34) H. C. Honsell, *Uber den naturlichen Strombau des Oberrheins* (V.D.G. 7, Berlin 1887). O. Baschin, *Die Entstehung der Flussmäander* (P.M. 1916, 62).

page 25

(35) G. Wagner, *Die Landschaftsformen von Würtembergisch-Franken* (E.L.A. 1919, 1).
(36) A. C. Ramsey, *The Physical Geology and Geography of Great Britain* (3rd Ed., London 1872, p. 243).
(37) W. M. Davis, *The Development of River Meanders* (G.M. 1903, 10). W. Behrmann, *Die Formen der Tieflandsflusse* (G.Z. 1915, 21).
(38) G. Wagner, *see* translator's note 35. H. Bach, *Die Theorie der Bergzeichnung in Verbindung mit Geognosie, etc.* (Stuttgart 1853).

Chapter III, page 27

(39) S. Passarge, *Physiologische Morphologie* (M.G.H. 1912, XXVI, p. 327).
(40) L. Rütimeyer, *see* translator's note 9.

page 28

(41) C. E. Dutton, *Tertiary History of the Grand Canyon Region* (U.S. Geogr. Geol. Survey monograph 2, Washington D.C. 1882, 264 pp. with Atlas). *Idem, The physical history of the Grand Canyon district* (A.R.G.S. 1882, 2, pp 47–166). W. M. Davis, *An excursion to the Grand Canyon of the Colorado* (B.M.C.Z. 1901, 38, 107–201). A. Heim, *Die Geschichte des Zürichsees* (N.N.G.Z. 1891). E. Brückner, in A. Penck and E. Brückner, *Die Alpen im Eiszeitalter* (Leipzig 1901–1909).

page 29

(42) E. Richter, *Geomorphologische Untersuchungen in den Hochalpen* (P.M. (erg-h) 132, 1900).

page 31

(43) L. Rütimeyer, *see* translator's note 9. A. Heim, *see* translator's note 32. A. Bodmer, *Terrassen und Talstufen in der Schweiz* (Dissertation, Zürich 1880). E. Brückner, *see* translator's note 41.
(44) E. Richter, *see* translator's note 42.
(45) J. Sölch, *Epigenetische Erosion und Denudation* (G.R. 1918, 9). L. Rütimeyer, *see* translator's note 9.

page 32

(46) A. Penck, *Die Vergletscherung der deutschen Alpen* (Leipzig 1882).
(47) W. Sievers, *Die heutige und fruhere Vergletscherung Sud-Amerikas* (V.G.D.N. 1911, 83).

page 34

(48) *See* especially Ch. VI, pp. 70–8, and translator's note 12.
(49) W. Klüpfels, *Uber Reliefmorphologie und zyklische Landschaftsgenerationen* (G.R. 1926, 17).

Chapter IV, page 35

(50) C. Lyell, *On the recession of the falls of Niagara* (P.G.A. 1842, 3, pp. 595–602). H. T. de la Beche, *A Geological Manual* (2nd Ed., London 1832). L. Rütimeyer, *see* translator's note 9. A. Heim, *see* translator's note 32. T. Kjerulf, *Die Geologie des sudlichen und mittleren Norwegen* (Bonn 1880).
(51) G. de Geer, *A geochronology of the last 12,000 years* (C.I.G.E. 1910, 241–53).

page 36

(52) N. Desmarest, *see* translator's note 136.

page 37

(53) A. Heim, *Mechanismus der Gebirgsbildung* (Atlas, Basle 1878). G. K. Gilbert, *see* translator's note 29. E. Suess, *The Face of the Earth* (Oxford 1904, p. 146 f.). E. Reyer, *Die Euganeen: Bau und Geschichte eines Vulkans* (Vienna 1877).

page 48

(54) A. Hettner, *Gebirgsbau und Oberflächengestaltung der Sächsischen Schweiz* (F.L.V. 1887, 11).

page 50

(55) G. K. Gilbert, *Harriman Alaska Expedition: Alaska III Glaciers and Glaciation* (New York 1904).
(56) A. C. Ramsey, *The Excavation of the Valleys of the Alps* (P.M.L. 1862, 4th series, 24, 377–80). *Idem, On the Glacial Origin of Certain Lakes in Switzerland, the Black Forest, etc.* (Q.J.G.S. 1862. 18, 185–204). J. Tyndall, *The Glaciers of the Alps* (London 1860). *Idem, On the Conformation of the Alps* (P.M.L. 1862, 24). A. Penck, *see* translator's note 46. A. Heim, *Handbuch der Gletscherkunde* (Stuttgart 1885).

Chapter V, page 52

(57) W. Deecke, *see* translator's note 28.

page 53

(58) C. W. von Gümbel, *Geognostische Beschreibung des Konigreichs Bayern* (Kassel 1891).
(59) A. Penck, *Die Bildung der Durchbruchstäler* (S.V.N.K. 1888).

page 54

(60) A. C. Ramsey, *On the Denudation of South Wales* (M.G.S.B. 1846, 1). F. von Richthofen, see translator's note 29 (pp. 173 ff.). J. W. Powell, *Report on the geology of the eastern portion of the Uinta Mts* (Washington, 1876).
(61) J. B. Jukes, see translator's note 27. W. Topley, *The Geology of the Weald* (M.G.S.B. 1875).

page 55

(62) G. Bischof, *Elements of Chemical and Physical Geology* (T.C.S. 3 vols, London 1854–9). C. Lyell, *The Geological Evidence of the Antiquity of Man* (London 1863). H. B. Medlicott, *On the Geological Structure and Relations of the southern portion of the Himalayan Range between the Rivers Ganges and Ravee* (M.G.S.I. 1860, 3). E. Tietze, *Einige Bemerkungen über die Bildung von Quertälern* (J.G.R.A. 1878).
(63) J. B. Jukes, see translator's note 27. R. Rütimeyer, see translator's note 9. A. Heim, see translator's note 32. G. K. Gilbert, see translator's note 29. F. Löwl, *Die Entstehung de Durchbruchstäler* (P.M. 1882, XXVIII, 405–416).

page 56

(64) O. Krümmel, *Einseitige Erosion* (Ausland 1882).

page 59

(65) E. Richter, see translator's note 42.

Chapter VI, page 60

(66) *See* author's note 35 and translator's note 12.

page 62

(67) G. Wagner, *Uber das Zuruckweichen der Stufenrander im Schwaben und Franken* (J.O.G.V. 1924, 13). See also translator's note 35.
(68) E. Tietze, *Bemerkungen über die Bildung der Querthäler* (J.G.R.A. 1882, XXXII, pp. 685 ff.).
(69) J. W. Powell, see translator's note 32. C. E. Dutton, see translator's note 41.
(70) A. Hettner, see translator's note 54.

page 63

(71) W. M. Davis, see translator's note 41.
(72) H. Rassmuss, H. von Staff, *Zur Morphologie der Sächsischen*

Schweiz (G.R. 1911, 2). H. von Staff, *Zur Morphogenie der Präglazial Landschaft der Westschweizer Alpen* (Z.D.G.G. 1912, 64).
(73) R. Gradmann, *Das Schichtstufenland* (Z.G.E. 1919, 54).
(74) G. K. Gilbert, see translator's note 29, and likewise for F. von Richthofen and A. Philippson. J. Sölch, *Eine Frage der Talbildung* (Penck-Festschrift, Stuttgart 1918). *Idem, Ungleichseitige Flussgebiete und Talquerschnitte* (P.M. 1918, 64).

page 65
(75) See geological table, p. 166.
(76) The Rigi is a 1787-metre eminence of the Alps of southern Germany. See L. Rütimeyer, *Der Rigiberg* (Basle 1877).

page 67
(77) M. Neumayr, see author's note 50. A. Penck, see author's note 62. E. Richter, see translator's note 42.
(78) H. Schmitthenner, *Die Entstehung der Stufenlandschaft* (G.Z. 1920, 26). H. Weber, *Die Oberflächenformen der Tambacher Schluchte bei Eisernach* (F.L.V. 1926, 24). E. Philippi, *Über die präoligozane Landoberfläche in Thuringen* (Z.D.G.G. 1910, 62).

page 68
(79) T. Fischer, *Die Abrasionsküste bei Tipaza und Algier* (P.M. 1885, 31, and 1887, 33, as well as in the Z.G.E. 1906, 42).
(80) H. W. Ahlmann, *Mechanische Verwitterung und Abrasion an der Grundgebirgsküste des nordwestlichen Schonen* (B.G.I.U. 1916, 13). *Idem, Geomorphological Studies in Norway* (G. AN. 1919, 1).

page 70
(81) W. Bornhardt, *Zur Oberflachengestaltung und Geologie Deutsch Ostafrikas* (Berlin 1900). S. Passarge, *Die Kalahari* (Berlin 1904). *Idem, Die Inselberglandschaft der Masaisteppe* (P.M. 1923, 69). F. Jaeger, see author's note 13. *Idem, Geographische Forschungen im abflusslosen Gebiet von Deutsch-Ostafrika* (Innsbruck 1912). F. Thorbecke, *Die Inselberglandschaft von Nord Tikar* (Hettner Festschrift, Breslau 1921). E. Obst, *Die Masaisteppe und das Inselbergproblem* (M.G.H. 1913, XXVII). L. Waibel, *Die Inselberglandschaft von Arizona und Sonora* (Z.G.E. Sonderband 1928). H. Schmitthenner, see author's note 13. In citing O. Maull, Hettner may be referring to that author's *Vom Itatiaya zum Paraguay*, although the work was not published in Leipzig until 1930.
(82) B. Brandt, *Die Tallosen Berge an der Bucht von Rio de Janeiro* (M.G.H. 1917, XXX).

page 72
(83) A. de Lapparent, *La question des pénéplains envisagée à la lumière des faits géologiques* (C.I.G. VII, Berlin 1900, 11).

page 74

(84) E. Brückner, in A.S.P.N. 102, XIV. F. Machatschek, *Verebnungsflächen und jungere Krustenbewegungen in Alpinen Gebirgssystem* (Z.G.E. 1916, 51). And idem, *Der Schweizer Jura, Versuch einer geomorphologischen Monographie* (P.M. (Erg-h) 1905, 150).

page 76

(85) J. Sölch, see translator's note 74, in *Grundfragen der Landformung in den nordöstlichen Alpen* (G. AN. 1922, 4) and in *Das Formenbild der Alpen* (G.Z. 1925, 31). W. Penck, see author's note 1.

(86) S. Passarge, *Die Vorzeitformen der deutschen Mittelgebirgs Landschaften* (P.M. 1919, 65).

page 77

(87) R. Chudeau, *Sahara Soudenais* (Paris 1909).

page 78

(88) E. Philippi, see translator's note 79.

(89) O. Maull, *Die Germanische Rumpffläche als Arbeitshypothese* (ANZ 1921). N. Krebs, *Eine Karte der Reliefenergie Suddeutschlands* (P.M. 1922, 68).

(90) W. Penck, *Die Piedmontflächen des sudlichen Schwarzwaldes* (Z.G.E. 1925, 60). H. Schrepfer, *Oberflächengestalt und eiszeitliche Vergletscherung im Hochschwarzwald* (G. ANZ. 1926).

page 79

(91) E. de Martonne, *Recherches sur l'evolution morphologique des Alpes de Transylvanie* (R.G.A. 1907). E. Brückner, see translator's note (85) H. von Staff, see translator's note 72. K. Ostreich, *Himalayastudien* (Z.G.E. 1914, 49). A. Philippson, *Reisen und Forschungen im westlichen Kleinasien* (P.M. (Erg-h) 1910, 167; 1911, 172; 1913, 177; 1914, 180; 1915, 183). B. Willis, *Research in China* (Carnegie Institute Publication 54, Washington 1907).

page 80

(92) H. Weber, see translator's note 79.
(93) E. Brückner, see translator's note 85.

page 81

(94) E. Richter, see translator's note 42.

(95) A. Briquet, see author's note 59. E. Philippi, see translator's note 79. H. von Staff, see translator's note 72.

page 82

(96) H. Schmitthenner, *Die Suddeutschen Stufenlandschaft und der Graben der Rheinbene* (B.O.L. 1927).

(97) L. Rütimeyer, see translator's note 9. A. Heim, *Uber Talstufen und Terrassen in den Alpentälern* (V.N.G. 1878, 23).

page 83
(98) G. Braun, *Beiträge zur Morphologie des nordlichen Apennin* (G.Z.E. 1907, 43).
(99) A. Penck, see translator's note 46.
(100) H. von Staff, see translator's note 72.
(101) A. Philippson, *Grundzüge der allgemeinen Geographie* (2 vols., Leipzig 1923/24).

Chapter VII, page 87
(102) J. Partsch, *Schlesien, eine Landeskunde* (Breslau 1896–1911).
(103) H. Spethmann, *Härtling für Monadnock—Nachrumpf und Vorrumpf* (Z.M.G.P. 1908, pp. 746–8).

page 88
(104) E. Richter, *Die Gletscher der Ostalpen* (Stuttgart 1888).

page 89
(105) S. Passarge, *Verwitterung und Abtragung* (V.D.G. Lubeck 1910).

page 90
(106) J. Partsch, *Die Vergletscherung der Riesengebirges zur Eiszeit* (F.L.V. 1894, 8).

Chapter VIII, page 92
(107) B. Cotta, see translator's note 20.

page 93
(108) F. Löwl, see translator's note 63. Idem, *Uber Talbildung* (P.M. 1882, 28).

page 98
(109) E. Suess, see translator's note 53.

Chapter IX, page 102
(110) G. K. Gilbert, *Lake Bonneville* (U.S. Geogr. Geol. Survey Monograph 1, Washington D.C. 1890).

page 103
(111) A. von Böhm, *Abplattung und Gebirgsbildung* (Leipzig/Vienna 1910).

page 104
(112) W. Behrmann, *Die Oberflächengestaltung des Harzes* (F.L.V. 1912, XX).

page 108

(113) E. F. Gautier, *La Sahara* (Paris 1928).

page 109

(114) F. von Richthofen, *On the mode of origin of the loess* (G.M. 1882, 9, 293–305).
(115) W. A. Obrutschev, *Zur Frage der Entstehung des Loess* (Tomsk 1911).

page 110

(116) R. Lucerna, *Die Trogfrage* (Z.G.L. 1910/11, 5).

page 111

(117) O. Frass, *Aus dem Orient; Geologische Beobauchtungen am Nil, auf der Sinai Halbinsel und in Syrien* (Stuttgart 1867). E. Hull, *The Survey of Western Palestine* (London 1886). G. K. Gilbert, *see* translator's note 110.
(118) M. Blanckenhorn, *Neues zur Geologie und Palaeontologie Ägyptens* (Z.D.G.G. 1900, 52; 1901, 53), and *Geologie Ägyptens* (Berlin 1901).

page 112

(119) S. Passarge, *Die Kalahari* (Berlin 1904).
(120) G. K. Gilbert, *see* translator's note 110. J. C. Russell, *Subaerial deposits of the arid regions of North America* (G.M. 1889, 6).
(121) R. Lang, *Verwitterung und Bodenbildung als Einführung in der Bodenbildung* (Stuttgart 1920).
(122) O. Heer, *Die Fossile Flora Grönlands* (Leipzig 1883).
(123) A. Penck, *Die Formen der Landoberfläche und die Verschiebungen der Klimagürtel* (S.P.A.W. 1913, XLVI).

page 113

(124) J. Walther, *see* author's note 12.
(125) S. Passarge, *see* translator's note 81.
(126) W. Köppen, A. Wegener, *Die Klimate der geologischen Vorzeit* (Berlin 1924).
(127) A. Nehring, *The Fauna of central Europe during the period of the Loess* (G.M. 1883, 10, 51–8).

Chapter XI, page 125

(128) E. Suess, *see* translator's note 53.
(129) O. Peschel, *Neue Probleme der Vergleichenden Erdkunde* (3rd Ed., Leipzig 1878, pp. 9–23).

page 126

(130) E. Réclus, *La Terre* (Paris, 1866–7).

(131) R. Credner, *Die Deltas* (P.M. (Erg-h) 1878, 56).
(132) L. Rütimeyer, *Die Bretagne* (Basle 1883).

page 127

(133) P. Gulliver, *Shoreline Topography* (P.A.A.S. Boston 1899).

Chapter XII, page 130

(134) J. E. Guettard, *On the degradation of mountains effected in our time by heavy rains, rivers and the sea* (Mémoires sur différentes parties des Sciences et Arts, 5 vols., Paris 1768–83). N. Desmarest, *Mémoire sur clef de 3 epoques de la nature par les produits des volcans* (Memoire de L'Institute, Tome VI, 1806). J. Hutton and likewise J. Playfair, *see* translator's note 27. J. L. Heim, *Geologischer Versuch über die Bildung der Täler durch Ströme* (Weimar 1791). K. E. A. von Hoff, *see* translator's note 16. C. Lyell, *Principles of Geology* (1st Ed., 3 vols., London 1830–33).

page 133

(135) *See* translator's note 25.
(136) A. Philippson, *Das Mittelmeergebiet* (3rd Ed., Leipzig/Berlin 1914, Chaps. V and VI, pp. 89–142).

Chapter XIII, page 137

(137) C. von Songklar, *Allgemeine Orographie, die Lehre von den Reliefformen der Erdoberfläche* (Vienna 1873). A. Neuber, *Charakteristik und Terminologie der Bodengestalt der Erdoberfläche* (Vienna 1901).

page 139

(138) *See* translator's note 5.

page 142

(139) E. Brückner, *see* translator's note 84. G. Braun, *see* author's note 58.
(140) H. von Staff, *see* translator's note 72.

Appendix, page 148

(141) R. Credner, *see* translator's note 131. *Idem, Die Reliktenseen* (P.M. (Erh-h) 1887–88, 86 and 89).

page 149

(142) A. Daubrée, *Synthetische Studien zur Experimentalgeologie* (Braunschweig 1880).

page 150

(143) A. Philippson, *Studien über Wasserscheiden* (M.V.E.L. 1886).

(144) W. M. Davis, *Die Erklärende Beschreibung der Landformen* (Leipzig 1912, p. xvi and p. 73, as well as at C.I.G. 9 Geneva 1908).

page 157

(145) W. Brancá, *Schwabens 125 Vulkanembryonen* (Stuttgart 1894).

page 160

(146) A. von Humboldt, *Versuch die mittlere Hohe der Kontinente zu bestimmen* (A.P.C. 1842, 57).
(147) C. von Songklar, *see* translator's note 137.

page 161

(148) S. Passarge, *Morphologie des Messtischblatts Stadtremda* (M.G.H. 1914, XXVIII).

page 162

(149) G. Götzinger, *see* translator's note 17.
(150) H. Gehne, *Eine neue methode Geomorphologischer Kartendarstellung* (P.M. 1912, 58). W. Behrmann, *see* translator's note 112. S. Passarge, *see* translator's note 148. F. Machatschek, *see* author's note 93. C. Rathjens, senr., *Morphologie des Messtischblatts Saalfeld* (Hamburg 1920).

page 164

(151) A. Stübel, *Die Vulkanberge von Ecuador* (Berlin 1897).
(152) W. M. Davis, K. Östreich, *Praktische Ubungen in physischer Geographie* (Berlin/Leipzig 1918).

Index

Actualism 6
Age, concept of 36, 97–9
  Davis's interpretation of 37
  of benchlands 66
  of valleys 35–51, 57
  tectonic 97

Base-level of erosion 69
Benchlands 61–7
Blanckenhorn stresses 111
Block diagram 164
'Boxed' landscape 34

Canyons 47, 48
Chasms 48
Climate 33
  and material relocation 116
  minor landform dependence on 10
Climatic changes 143
Climatic development 102, 107–14
Cloudbursts 122
Coasts and coastline processes 125–128
Comparative method 45
Corries 68
Creep 7
Cuestas 61
Cycle theory 72, 103, 134–6

Dating, geological 35
Deductive method xxx, 58, 150–3
Degradation 63, 71, 105
Dells 75
Demonstration of causes 155
Denudation 26, 86, 93, 144
Denudation terraces 28, 33–4
Descriptions 3, 160
Desert valleys 91
Deserts 5, 10, 50, 90, 111, 123, 133

Drainage pattern 43, 56, 58, 82
Drainage systems 58

Eminences 87
Epeirogenic movement 102, 106
Erosion 7, 26, 139
  base-level of 69
  direct, indirect, vertical and lateral 20
  theory of 14, 18, 150
Erosion cycle 105
Erosion terraces 29, 31–4
Erosional nature of valleys 14
Escarpments 61
Evaporation 120
Explanation 155

Firn 117
Fissure nature of valleys 14, 17
Fjords 124, 126
Fluvial transference of material 119–123
Forces, action and interaction of 5
  sub-aerial 6
Formative processes 6

Genetic terminology 156, 159
Geological table for South Germany 166
Geology and geologists xxiv
Geomorphological interrelationship of landscapes 115–23
Geomorphology, and its media xxvi
  and rock formation 93–7
  classification 137
  development xxix, 124
  geologically-oriented xxv
  importance of xxiii
  schools of thought in xxxi

Geomorphology—*contd.*
  science of 130
  scientific development of 130-4
Glaciers 32, 47, 50, 90, 110, 117, 118, 122
Gravel terraces 30, 32-4

'Hogs' backs' 65
Horsts 157

Illustrations 3, 163
Inductive method of research 150
Inselbergs 70
Intensity of dissection 87
Internal build xxiv, xxix, xxx, 26, 43, 57, 85, 132, 141, 144
  and surface form, discordance between 81
  formation of 100
  land surface dependence on 92-9
  use of term 95
Investigation methods 3

Karst regions 69

Land appraisal xxvi
Land surface, dependence on surface build 92-9
  development of 100-14
  subdivision and grouping of 145-7
  theories on origin of 129-36
Landform assemblages 137-44
Landforms, analysis of 5
  causes of xxviii
  coastal 124-8
  dependence on rock-type and disposition 8
  distribution of 4, 59
  minor 1-5, 10, 11
  origin of 3
  science of xxvi
  terminology 157
Landscape, 'boxed' 34
  geomorphological characterisation of 141
  geomorphological interrelationship of 115-23
  minor features of 1-13
  morphological character of 140-3
  structural style 140
Last Glaciation 12

Maps xxvi, 45, 95, 160
  comparative study method of research 148
  geological 161
  geomorphological 163
  morphological 161-3
  topographical 162
Marine abrasion 72, 75
Marine degradation 71
Mass-movement 6, 7
Mass-wasting 7
Material movement processes 115
Material relocation 115, 144
Material transference, fluvial 119-23
Minor landforms. *See* Landforms, minor
Monadnocks 87
Moraines 118
Morphometry 160
Mountains 96-8, 131, 135, 144
  origin of 85
  structural plan 85-8
  structural style 88-91

Nappe theory 131
Neckar valley 138
Neptunist theory 129

Oldlands 86, 87, 89
Ontological method 13
Orometry 160-1

Paralometry 160
'Pass', use of term 157
Peneplains and peneplain theory 60, 62-3, 68, 72, 75, 78, 79, 81, 83, 84, 105, 152
Photography 3, 163
Pictorial views xxvii
Piedmont flats 74
Piedmont stairways 74, 75
Piedmont surface 75
Plains 62
Planations 60, 65, 68-70, 75, 76, 106
Plateaux 55, 64, 89
Pliocene 74
Plutonist theory 129
Presentation methods 153-6
Profile of equilibrium 22
Pseudo-valleys 50

Rainfall 120

Index

Ravines 47
Redeposition 144
Regolith, study of 4
Relief energy 87, 89
Relief models xxvi
Relocation 6
Remnant oldlands 142–3
Remnant plain 73
Remnant surfaces 60, 70–7, 80–4
Research methods 148–53
Rias 124
River meanders 25, 82
Rivers 16, 119–23
Rock formation and geomorphology 93–7
Rock hardness 56
Rock permeability 75
Rock pillars 91
Rock platforms 64
Rock resistance 75, 88
Rock type 8, 63, 92–3
Rock weakness or hardness 93
Rock zones 56
Rocks, geological classification of 94

Saddles 88
Scarplands 61, 63, 65, 66
Sheet-flooding 70
Sheet-wash 8
Sketches 164
Soil thrust 7
Steppes 133
Strandflats 68
Strata, deposition of 60
Sub-aerial degradation 71
Sub-aerial planation 75, 76
Summits 80–1, 87, 88
Surface modification 139, 143, 141, 144

Tablelands 86, 87, 89
Tectonic age 97
Tectonic development 102–7
Tectonic features 140

Tectonic surface 58, 73, 85, 96, 107, 132, 141
Tectonic terms 157
Tectonics xxv, xxix, 95, 131
Terminology 156–60
Transportation 6, 144
Tropics 133
Truncating surfaces 60, 62, 67

Uplift 26, 87, 105–7

Valley bottom 30, 41–2, 49
Valley floor 31, 32
Valley meanders 25
Valley terraces 27–34, 49, 104
Valleys 86, 96–7, 103, 104, 135
  age and form of 35–51, 57
  alignment and arrangement of 52–9
  consonant or concordant 53
  desert 91
  erosion 14, 18, 20
  fissure nature of 14, 17
  flat-floored 49
  glacial 50
  hanging tributary 50
  inconsonant or discordant 53
  notch 49
  origin of 14–27, 52
  profile of equilibrium 22
  pseudo 50
  subsequent 55
  surviving 55
  transverse 56
  U-shaped 49, 50
  use of term 17
  V-shaped 49
Views 163
Volcanoes 85, 104, 107

Wadis 50, 91
Watershed residuals 87
Weathering 6–7, 86, 93, 115, 144
Weathering terraces 28
Wind 5, 7, 122